# Hyogo-Ken Nanbu Earthquake of January 17, 1995:

## A Post-Earthquake Reconnaissance of Port Facilities

Committee on Ports and Harbors Lifelines of the
Technical Council on Lifeline Earthquake Engineering of the
American Society of Civil Engineers

Edited by Stuart D. Werner, Dames & Moore, Earthquake Engineering Group, San Francisco, CA
and
Stephen E. Dickenson, Oregon State University, Department of Civil Engineering, Corvallis, OR

Published by the
American Society of Civil Engineers
345 East 47th Street
New York, New York 10017-2398.

ABSTRACT:

The objective of this committee report, *Hyogo-Ken Nanbu Earthquake of January 17, 1995: A Post-Earthquake Reconnaissance of Port Facilities*, was to observe and evaluate the seismic performance of ports in the Osaka Bay region of Japan. In addition to the actual observation and evaluation of the seismic performance of the port facilities, this scrutiny included numerous data-gathering meetings with representatives from cognizant port authorities, engineering consulting firms, construction companies, universities and private research organizations in Japan. The investigation was carried out over a 10-day period from February 18-27, 1995 and focused primarily on the seismic performance of the Port of Kobe, the Port of Osaka and, to a lesser degree, the Kansai Airport. This first-hand documentation of perishable data will enhance future engineering and research work at these and other ports around the world.

Library of Congress Cataloging-in-Publication Data

Dickenson, S. E.
Hyogo-Ken Nanbu earthquake of January 17, 1995 / Committee on Ports and Harbors Lifelines of the Technical Council on Lifeline Earthquake Engineering of the American Society of Civil Engineers ; edited by Stuart D. Werner and Stephen E. Dickenson.
p.   cm.
ISBN 0-7844-0161-6
1. Harbors—Earthquake effects—Japan-Osaka Bay Region.  2. Harbors—Earthquake effects—Japan—Hyogo-ken.  3. Earthquakes—Japan—Osaka Bay Region.  4. Earthquakes—Japan—Hyogo-ken.  I. Werner, Stuart D.  II. American Society of Civil Engineers. Committee on Ports and Harbors Lifelines.
TC306.H96D53  1996         96-6207
627'.3—dc20                CIP

   The material presented in this publication has been prepared in accordance with generally recognized engineering principles and practices, and is for general information only. This information should not be used without first securing competent advice with respect to its suitability for any general or specific application.
   The contents of this publication are not intended to be and should not be construed to be a standard of the American Society of Civil Engineers (ASCE) and are not intended for use as a reference in purchase specifications, contracts, regulations, statutes, or any other legal document.
   No reference made in this publication to any specific method, product, process or service constitutes or implies an endorsement, recommendation, or warranty thereof by ASCE.
   ASCE makes no representation or warranty of any kind, whether express or implied, concerning the accuracy, completeness, suitability or utility of any information, apparatus, product, or process discussed in this publication, and assumes no liability therefore.
   Anyone utilizing this information assumes all liability arising from such use, including but not limited to infringement of any patent or patents.

Photocopies. Authorization to photocopy material for internal or personal use under circumstances not falling within the fair use provisions of the Copyright Act is granted by ASCE to libraries and other users registered with the Copyright Clearance Center (CCC) Transactional Reporting Service, provided that the base fee of $4.00 per article plus $.50 per page is paid directly to CCC, 222 Rosewood Drive, Danvers, MA 01923. The identification for ASCE Books is 0-7844-0161-6/96  $4.00 + $.50 per page. Requests for special permission or bulk copying should be addressed to Permissions & Copyright Dept., ASCE.

Copyright © 1996 by the American Society of Civil Engineers,
All Rights Reserved.
Library of Congress Catalog Card No: 96-6207
ISBN 0-7844-0161-6
Manufactured in the United States of America.

Printed on recycled paper. 85% recovered fiber and 15% post-consumer waste.

# TABLE OF CONTENTS

                                                                             **Page**

ACKNOWLEDGEMENTS

INTRODUCTION .................................................................. 1-1
    1.1   Background ............................................................. 1-1
    1.2   Reconnaissance Objective and Scope ......................... 1-1
    1.3   Report Organization .................................................. 1-2

EARTHQUAKE OVERVIEW .................................................. 2-1
    2.1   Fault Rupture ........................................................... 2-1
    2.2   Ground Shaking ....................................................... 2-1
    2.3   Earthquake Effects ................................................... 2-2

KOBE PORT ....................................................................... 3-1
    3.1   General Background ................................................. 3-1
    3.2   Earthquake Ground Motions ...................................... 3-6
    3.3   Seismic Performance Overview .................................. 3-7
    3.4   Seismic Performance of Waterfront Retaining Structures and Cranes ....... 3-7
    3.5.  Seismic Performance of Pile-Supported Structures ...... 3-12
    3.6   Seismic Performance of Buildings ............................. 3-15
    3.7   Seismic Performance of Supporting Lifelines and Tanks ............ 3-16

OSAKA PORT ..................................................................... 4-1
    4.1   General Background ................................................. 4-1
    4.2   Ground Shaking ....................................................... 4-6
    4.3   Seismic Performance Overview .................................. 4-6

KANSAI AIRPORT ............................................................... 5-1
    5.1   General Background ................................................. 5-1
    5.2   Ground Shaking ....................................................... 5-3
    5.3   Seismic Performance Overview .................................. 5-3

CONCLUDING COMMENTS ................................................. 6-1

REFERENCES ..................................................................... R-1

# ACKNOWLEDGEMENTS

Funding for this reconnaissance was provided by the Port of Los Angeles, Port of Portland, Oregon Sea Grant Program, and Dames & Moore. This financial support is gratefully acknowledged.

The acquisition of background material on the port facilities in the Osaka Bay region, and the extensive logistical and travel arrangements required for this reconnaissance, would not have been possible without the generous assistance of numerous individuals in the United States and Japan. In particular, Dr. Ian Austin of Dames & Moore (Tokyo) and Mr. Kunio Arai of Chiyoda-Dames & Moore in Tokyo provided invaluable assistance with our meeting and reconnaissance arrangements, furnished important technical information on the effects of the earthquake, accompanied us during much of our field reconnaissance, and provided us with many of their excellent photographs (several of which are included in this report). In addition, valuable encouragement and post-earthquake data were provided by Mr. Douglas A. Thiessen of the Port of Los Angeles. Finally, we were greatly assisted by helpful discussions and information provided prior to our reconnaissance by Dr. Raymond B. Seed of the Department of Civil Engineering at the University of California at Berkeley, Mr. Michael Jordan and Mr. Feroze Vazifdar of Liftech Consultants, Inc. of Oakland, CA, and by Mr. Rinnosuke Kondoh of The International Association of Ports and Harbors, Tokyo, Japan.

While in Japan, numerous individuals generously gave of their time and expertise in meeting with us, providing us with substantial information on the port facilities and their seismic performance, accompanying us during portions of our reconnaissance, and assisting with arrangements and access to damaged port areas. We are particularly grateful to the following pre-eminent engineers, researchers, and port representatives for their valuable assistance along these lines:

| | |
|---|---|
| Dr. Hajime Tsuchida | Coastal Development Institute of Technology, Tokyo |
| Professor Masashi Kamon<br>Professor Kojiro Irikura | Disaster Prevention Research Institute<br>Kyoto University, Kyoto |
| Dr. Yoshinori Iwasaki<br>Dr. Takao Kagawa | Geo-Research Institute, Osaka |
| Mr. Nobuaki Yamamoto | Port & Harbor Bureau<br>Kobe City Government, Kobe |
| Mr. Tatsuo Wako | Third District, Port Construction Bureau<br>Ministry of Transport, Kobe |
| Mr. Kiyoshi Hatano<br>Mr. Nishii | Construction Division, Port & Harbor Bureau<br>City of Osaka |

| | |
|---|---|
| Mr. Mark Brady | International Relations, Port & Harbor Bureau<br>City of Osaka |
| Mr. Nobuyuki Kobayashi<br>Mr. Kazuya Tatsumi | Port Promotion Department, Port & Harbor Bureau<br>City of Osaka |
| Mr. Yoji Shimazu | Civil Works Marketing Division<br>Mitsui Construction Co., Ltd., Tokyo |
| Mr. Nobuaki Shiraishi<br>Mr. Yutaka Okuda | Engineering Division<br>Kansai International Airport Co., Ltd., Osaka |

Special recognition is also extended to Dr. Yuzuru Ito of the Japan Highway Public Corporation (Osaka Bureau) for his considerable efforts in arranging for our trip and providing us with a tour of highway structures effected by the earthquake. In addition, we are grateful to Mr. Sadayoshi Hikata (Civil Engineering Research Institute, Hokkaido Development Bureau) for generously translating documents, obtaining reference material, and assisting with our correspondence during his residency as a Visiting Scholar at Oregon State University. Finally, we would like to thank Mr. Demetrious Koutsoftas of Dames & Moore for his helpful and constructive review of this report.

# CHAPTER 1
# INTRODUCTION

## 1.1 Background

Past experience has shown that port facilities can be susceptible to severe damage from earthquake ground shaking and associated phenomena (e.g., ground deformation, liquefaction, submarine slope failures), and that such damage can result in significant economic losses to the port authority and to industries dependent on marine commerce. In recognition of this, the Ports and Harbors Committee of the ASCE-Technical Council for Lifeline Earthquake Engineering (TCLEE) is developing seismic guidelines for hazard evaluation, design, analysis, and emergency response/recovery for port waterfront, cargo handling/storage, and infrastructure components (Werner et. al., 1995). This Committee is comprised of experienced professionals from port authorities, consulting engineering firms, government, and universities.

An important source of information for these guidelines is the documented performance of ports during past earthquakes. Toward this end, the Committee has been actively compiling data on the performance of ports during historic earthquakes worldwide. It was felt that the Committee's data compilation efforts and overall seismic guidelines document would be greatly enhanced by an in-depth reconnaissance of ports in the region of moderate to strong ground shaking from the Hyogo-Ken Nanbu Earthquake. Furthermore, it was anticipated that the dissemination of this information to the ports community would foster an increased awareness of: (a) the potential vulnerability of ports to moderate to strong ground motions; (b) seismic performance characteristics of quay walls, fills, and other port components, and how this performance may be affected by the strength of the shaking and by design and fill placement procedures; (c) the potential risks to post-earthquake operations and repair/reconstruction efforts at ports due to damage to supporting utility and transportation lifelines; and (d) the potential effects of port damage on local, regional and international commerce and economies.

## 1.2 Reconnaissance Objective and Scope

The objective of this reconnaissance following the Hyogo-Ken Nanbu Earthquake was to observe and evaluate the seismic performance of ports in the Osaka Bay region of Japan (Fig. 1-1). The reconnaissance focused on first-hand documentation of perishable data for enhancing future engineering and research work at these and other ports in the United States and Japan, as well as the ASCE-TCLEE Ports and Harbors Committee's seismic guidelines.

This reconnaissance was carried out by Stuart D. Werner of Dames & Moore and Stephen E. Dickenson of Oregon State University, who are Chairman and Vice-Chairman, respectively, of the Ports and Harbors Committee. In addition to the actual observation and evaluation of the seismic performance of the port facilities, the reconnaissance included numerous data-gathering meetings with representatives from cognizant port authorities, engineering consulting firms, construction companies, universities and private research organizations in Japan.

The information gathered during our visit continues to be augmented with additional data provided by our colleagues in Japan and the United States who have also visited the ports after the earthquake, and also with data gleaned from the earthquake engineering literature from these countries. This information is being synthesized into reports that should contribute to an increased awareness in the ports community of the potential seismic risks to port facilities during moderate to large earthquakes. Relevant information and selected slides from this reconnaissance will be made available not only to the Ports and Harbors Committee, but also to the ports community with facilities located in regions of the United States that have a potential for damaging earthquakes.

The reconnaissance was carried out over a 10-day time period that extended from February 18-27, 1995. The investigations focused primarily on the seismic performance of the Port of Kobe and the Port of Osaka (Fig. 1-2). The reconnaissance was initiated with orientation and data-compilation meetings with persons in Tokyo, Kyoto and Osaka who have extensive experience with the ports. These meetings provided valuable information on seismic design procedures, construction practices, soil conditions, and soil improvement techniques employed at various portions of the port facilities. The data and guidance provided at this early stage of the reconnaissance enhanced subsequent field investigation efforts, which consisted of both walking and boat tours of the ports. Several of the tours were guided and arranged by port personnel, and numerous unaccompanied inspections were also conducted. In addition, we briefly met with personnel from the Kansai Airport, a major airport recently constructed on a 509 ha man-made island in Osaka Bay west of Osaka, in order to compile readily available information on the seismic performance of its island fills and quay wall components.

## 1.3 Report Organization

The remainder of this report is organized into five main chapters. Chapter 2 provides an overview of the seismological aspects of the Hyogo-Ken Nanbu Earthquake, and Chapters 3 through 5 describe our observations pertaining to the seismic performance of the Port of Kobe, the Port of Osaka, and the Kansai Airport. Chapter 6 contains concluding comments from the reconnaissance.

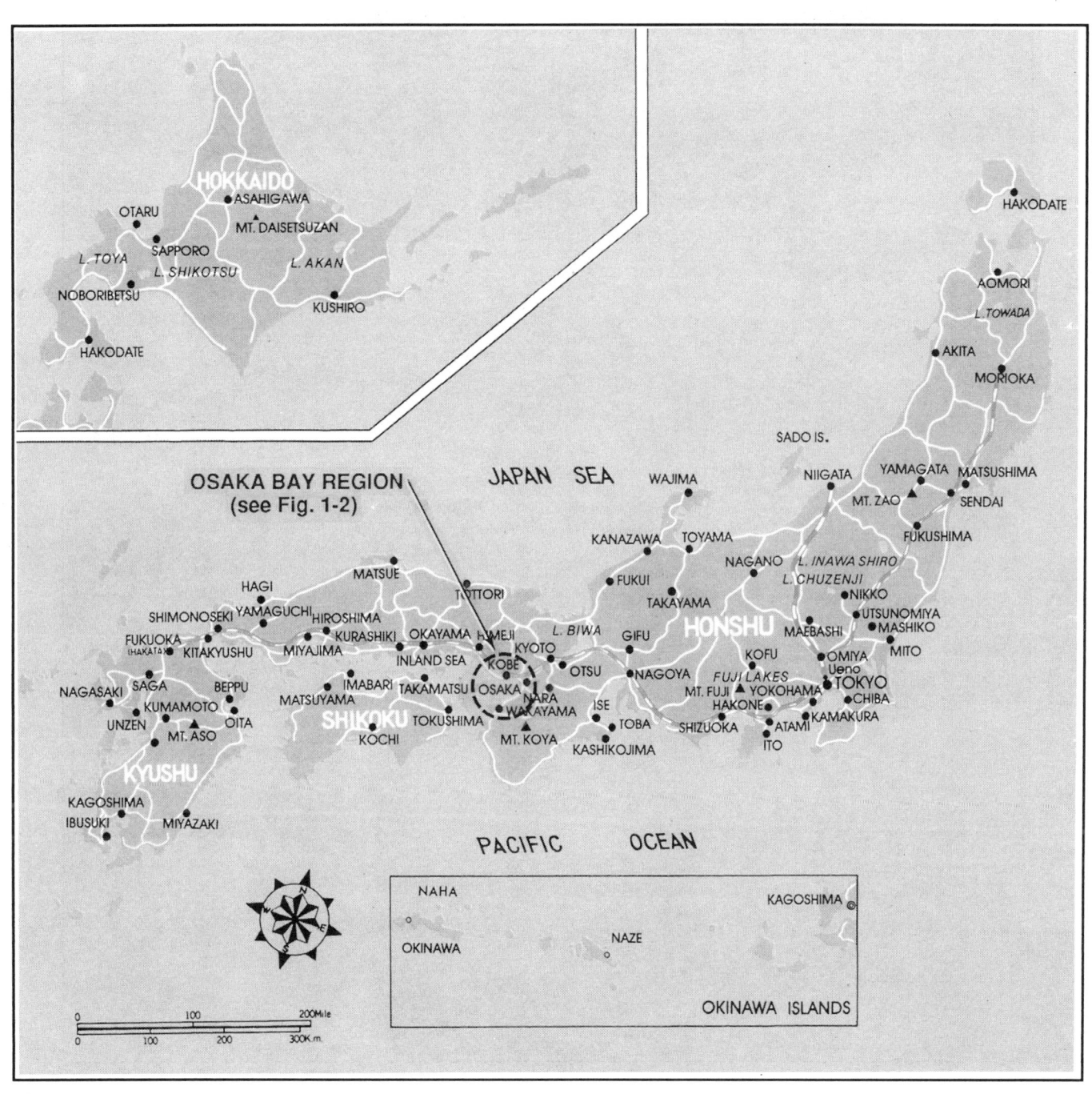

FIGURE 1-1
MAP OF JAPAN
(JTNO, 1986)

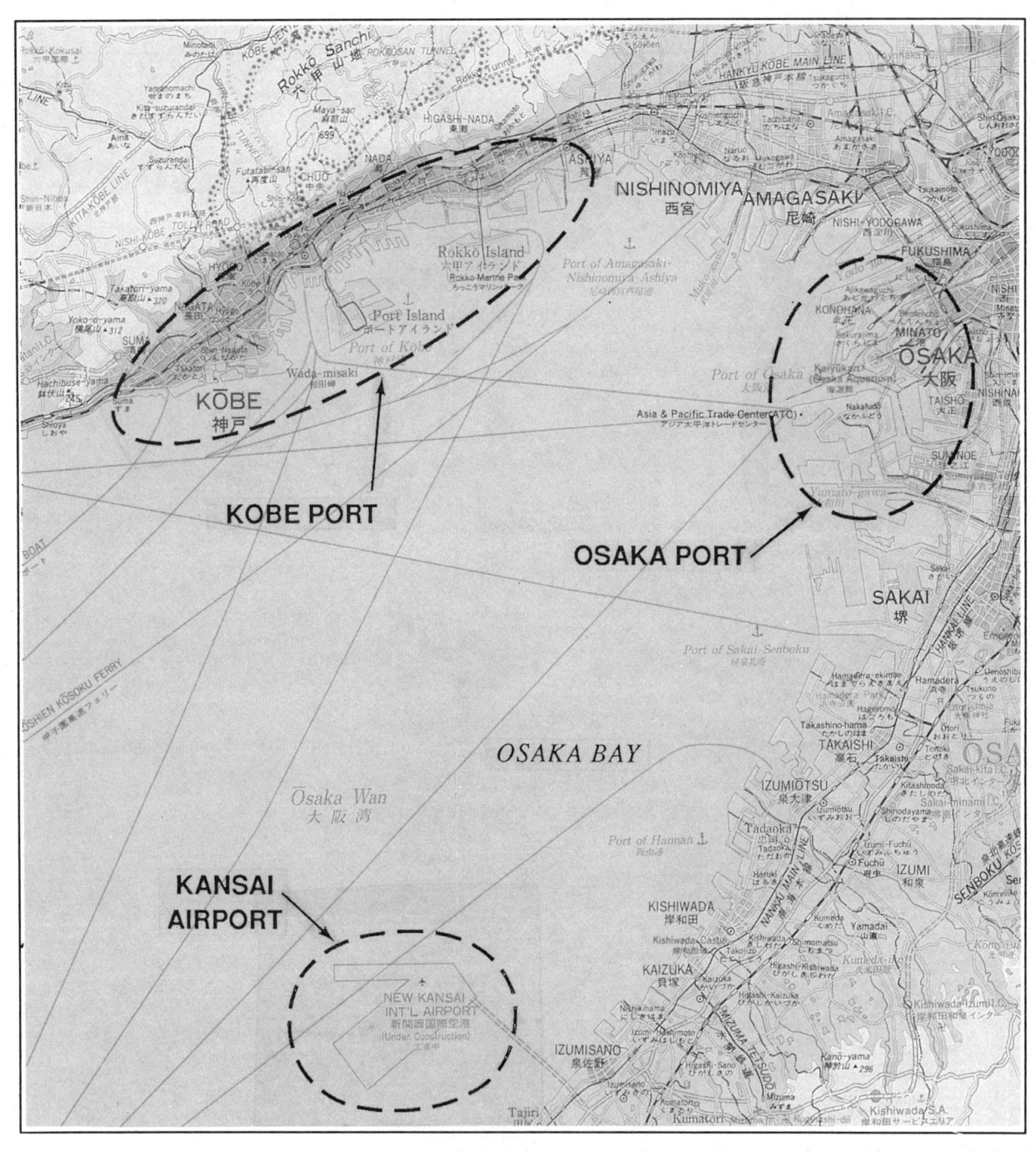

FIGURE 1-2
OSAKA BAY REGION
(EERI, 1995b)

# CHAPTER 2
# EARTHQUAKE OVERVIEW

## 2.1 Fault Rupture

The Hyogo-Ken Nanbu Earthquake ($M_w = 6.9$) occurred at 5:46 am (local time) on January 17, 1995. The rupture was initiated at a depth of approximately 10 km below the northeastern tip of Awaji Island (Figure 2-1a). In plan, this location is roughly 20 km southwest of downtown Kobe. Ground surface rupture has been mapped along the Nojima Fault located in the northwestern portion of Awaji Island. It is inferred from aftershock patterns (Figure 2-1b) that the strike-slip rupture propagated bi-laterally from the hypocenter, with the rupture to the east extending directly beneath the city of Kobe, a densely developed city with a population of approximately 1.5 million people. To the east of Awaji Island, a complex system of faults has been mapped on the alluvial plain between the base of the Rokko Mountains and the northern margins of Osaka Bay (RGAFJ, 1991). The general geometry of these features is shown in Figure 2-2. In this area, the rupture is surmised by many to have occurred along the Ashiya Fault. To date, no evidence of surface rupture has been reported in the Kobe area. The total rupture length has been estimated from aftershock patterns to have been about 30-to-50 km.

The bi-lateral mode of fault rupture experienced during this event is very similar to the rupture mechanism exhibited during the 1989 Loma Prieta Earthquake in the San Francisco Bay region. This type of faulting results in a duration of strong shaking that is about half of what would be considered characteristic for earthquakes of the same magnitude. The effects of this reduced duration were clearly overshadowed by the propagation of the fault rupture into the city of Kobe. This path of rupture propagation appears to have resulted in a directivity of seismic energy into the urban area of Kobe and into areas to the northeast, thereby enhancing the intensity of the ground motions adjacent to and northeast of the fault rupture (Somerville, 1995).

## 2.2 Ground Shaking

The combination of the directivity effects and the close proximity of the city of Kobe to the fault rupture resulted in extremely strong shaking, with peak horizontal ground accelerations recorded at strong motion accelerometer stations in Kobe that were often on the order of 500 to over 800 cm/sec$^2$ (Fig. 2-3). Peak vertical accelerations were generally about two-thirds of the peak horizontal accelerations. The duration of the strong shaking segment of these recorded motions was up to about 10 sec (Fig. 2-4). It is noted that the soil conditions at most of the accelerometer stations in Kobe consisted of alluvial deposits (predominantly older alluvial sediments in the downtown area, with a transition toward the bay to fill underlain by young marine deposits and by deep alluvium).

Along the margins of Osaka Bay and on the artificial islands of the Port of Kobe deep soil motions were modified by overlying strata of soft to medium stiff marine clay and loose sandy fill soils which have been placed since the late 1800s (Fig. 2-5). Strong motion records measured at the Port of Kobe (including records from a down-hole array on a site on Port Island where the surface fill layer liquefied) are described in Section 3.2.

Figure 2-6 shows a return period vs. peak acceleration (PGA) relationship for the Kobe area that was provided to us during our visit to Japan. This figure shows estimated return periods for PGAs of 500-800 cm/sec$^2$ that range from about 500 years to over 1000 years. These values are corroborated by the occurrence of an estimated Magnitude 7.25 earthquake in Kobe in 1596 (Tsukuda, 1987). It is interesting to note that an independent study made in Japan over 40 years ago estimated that peak accelerations on firm ground of approximately 190 cm/sec$^2$, 300 cm/sec$^2$ and 420 cm/sec$^2$ are associated with return periods of 75, 100 and 200 years, respectively, in the Kobe region (Kawasumi, 1951). Maps of the accelerations and return periods produced by Kawasumi are shown in Figure 2-7 and, for the Kobe area, agree reasonably well with the data in Figure 2-6. The return periods for these acceleration levels are comparable to those estimated for similar levels of ground accelerations in highly seismic regions in California.

Figure 2-8 compares peak horizontal accelerations and velocities recorded at soil sites in Kobe to those predicted for a strike-slip earthquake using empirical attenuation relationships for soil sites that are based mainly on California data (Somerville, 1995). This figure shows that the peak accelerations recorded in Kobe are generally comparable to those predicted by the empirical relationships, whereas the peak velocities recorded in Kobe tend to be somewhat larger. It is noted that the peak near-field velocities recorded in Kobe are comparable to the largest velocities recorded in Northern Los Angeles during the 1994 Northridge Earthquake (EERI, 1995a).

## 2.3 Earthquake Effects

The earthquake effects on the City of Kobe and in the immediately adjacent areas were often devastating. Over 5,300 people in the area were killed, nearly 27,000 people were injured, and an estimated 300,000 people were homeless after the earthquake. This was due primarily due to collapses of older houses, built of post and beam construction techniques, that had only minimal lateral force resistance and supported roofs covered with heavy clay tiles. Older engineered building structures also suffered major damage that included first-story and mid-story collapses, leaning, and severe shear and flexural damage. Major fires occurred in several areas of Kobe, and many highway and railroad bridges collapsed or were severely damaged (Fig. 2-9). Water, wastewater, and natural gas systems and components in the area were also severely damaged. The Kobe Port suffered extensive damage due to liquefaction of the uncompacted fills throughout the port, and due to quay walls that were inadequate to resist the increased lateral pressure that resulted from the associated pore water pressure buildups. Earthquake-induced losses in Kobe have been estimated to be as high as $200 billion, which is about an order-of-magnitude larger than the estimated losses in the Los Angeles area due to the 1994 Northridge Earthquake. Further extensive description of the effects of the Hyogo-Ken Nanbu Earthquake on the buildings and lifelines in the area is contained in several reports on the earthquake that have been produced in Japan and the United States in recent months (e.g., DPRI, 1995; INCEDE, 1995; EERI, 1995b; SEAONC, 1995).

It should be noted that although this reconnaissance focussed on the ports at Kobe and Osaka, several other ports in the region were affected by the Hyogo-Ken Nanbu Earthquake. The strong ground motions experienced along most of the margins of Osaka Bay and the northeastern portion of the Harina Sea (northwest of Awaji Island) exceeded 0.2 g (Fig. 2-3). The intensity of these ground motions corresponds to levels of shaking that have resulted in considerable damage to port facilities worldwide (Werner and Hung, 1982). An investigation of the seismic performance of other ports in the Osaka Bay area (e.g., Amagasaki-Nisinomiya-Ashiya, Sakai-Senboku, Hannan) and the waterfront regions northwest of Awaji Island (e.g., Takasago, Hirohata, and Abashi) is currently underway as a subsequent phase of this reconnaissance.

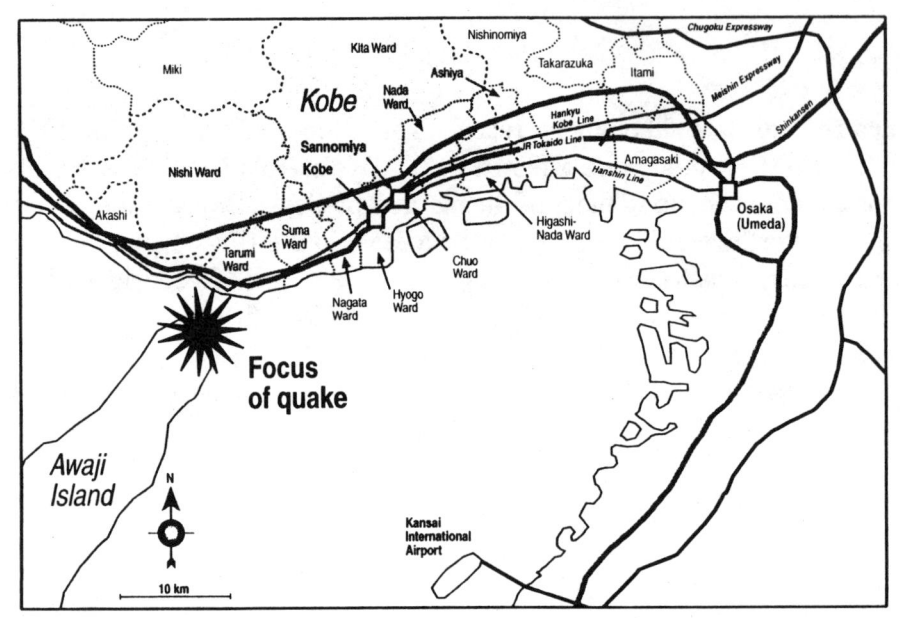

a) Main Shock on January 17, 1995 at 5:45 am (JTL, 1995)

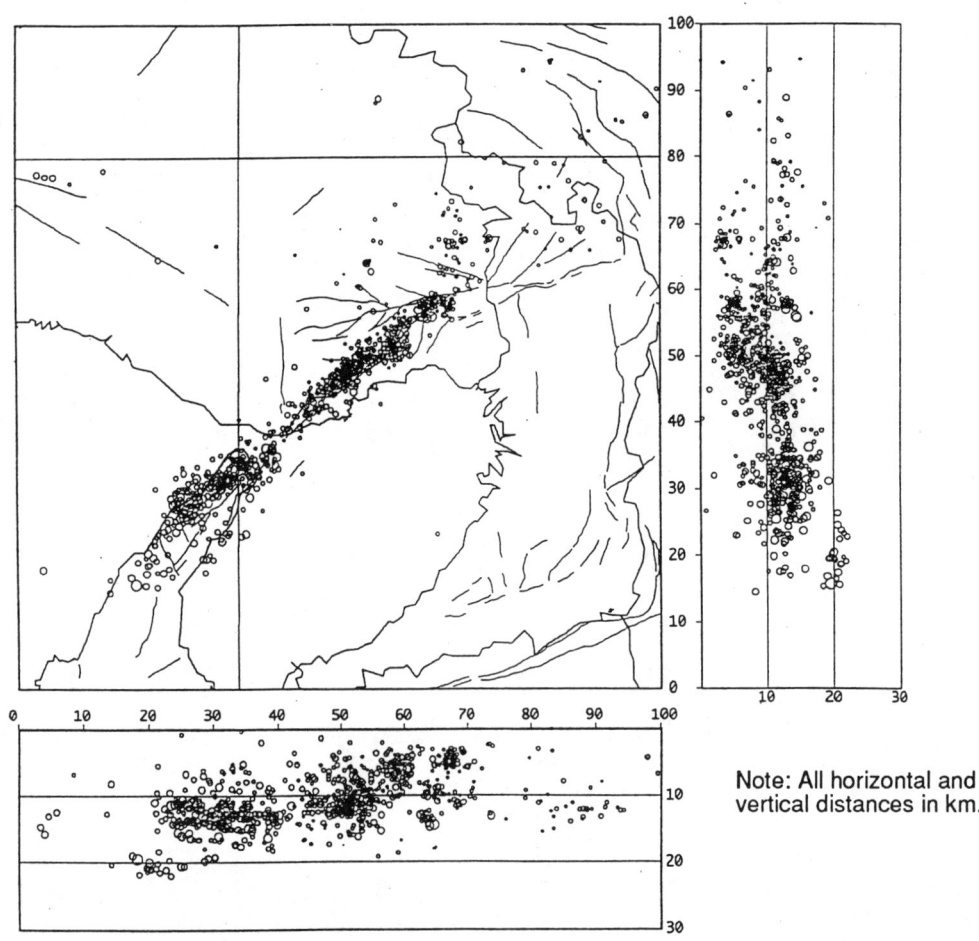

Note: All horizontal and vertical distances in km.

b) Aftershocks (1/19/95 to 1/27/95) (DPRI, 1995a)

**FIGURE 2-1
HYOGO-KEN NANBU EARTHQUAKE OF JANUARY 17, 1995**

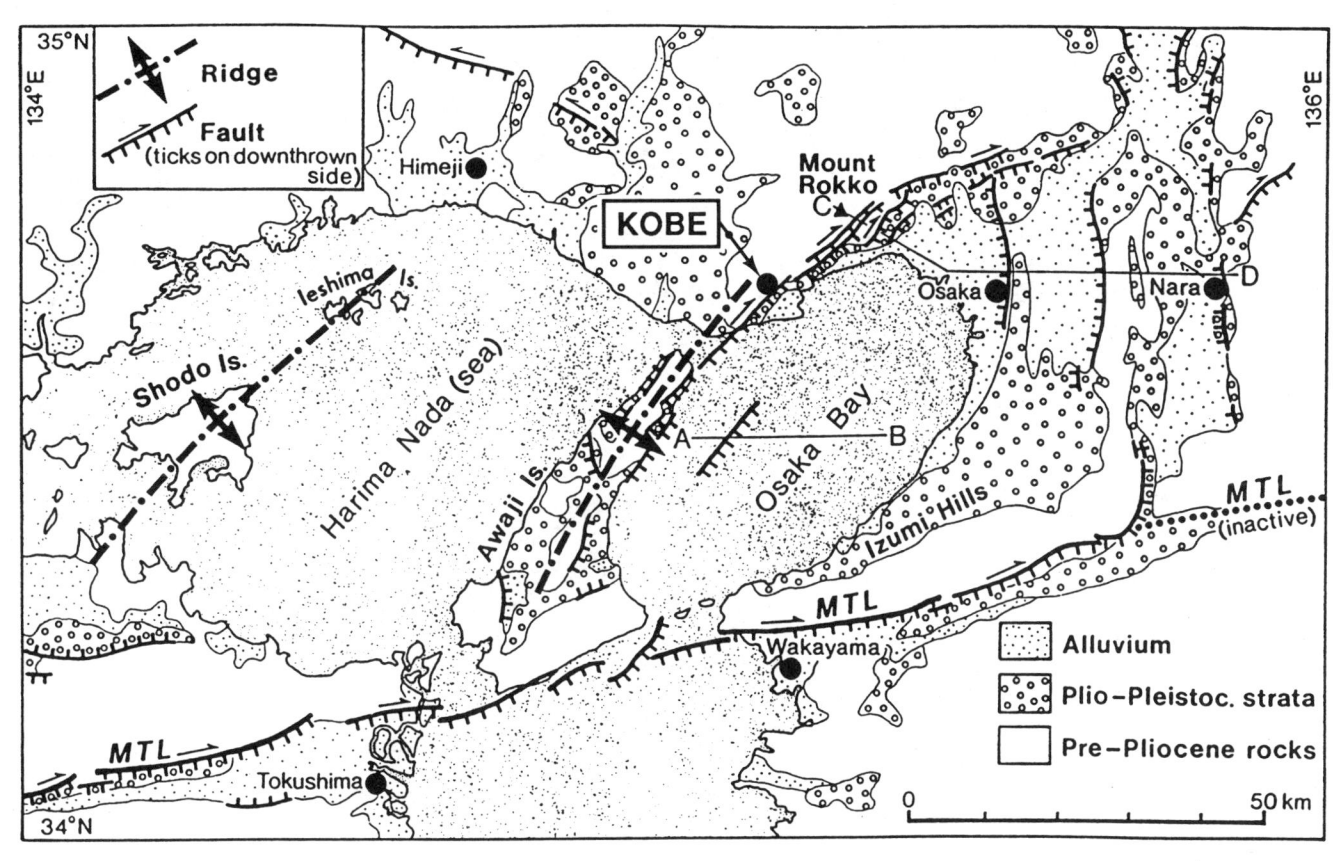

**FIGURE 2-2
REGIONAL TECTONIC MAP (Sugiyama, 1994)**

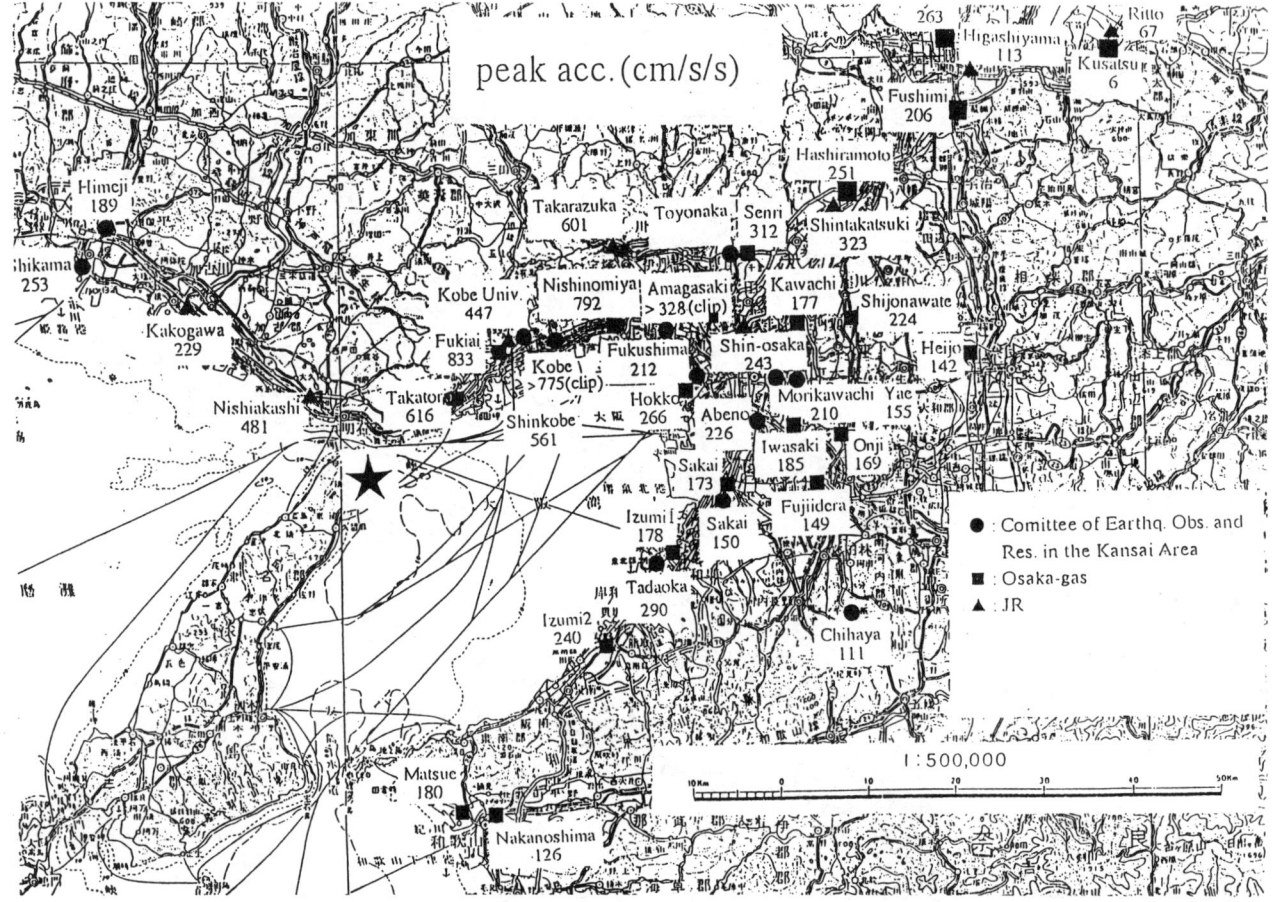

Peak Acceleration Measurements:

- Committee of Earthquake Observation and Research in the Kansai Area (Largest of Three Orthogonal Components)

- Osaka Gas (Vector Sum of Two Horizontal Components)

- JR (Vector Sum of Two Horizontal Components After They Have Been Lowpass Filtered at 5Hz)

**FIGURE 2-3
PEAK GROUND ACCELERATIONS
(GRI, 1995a)**

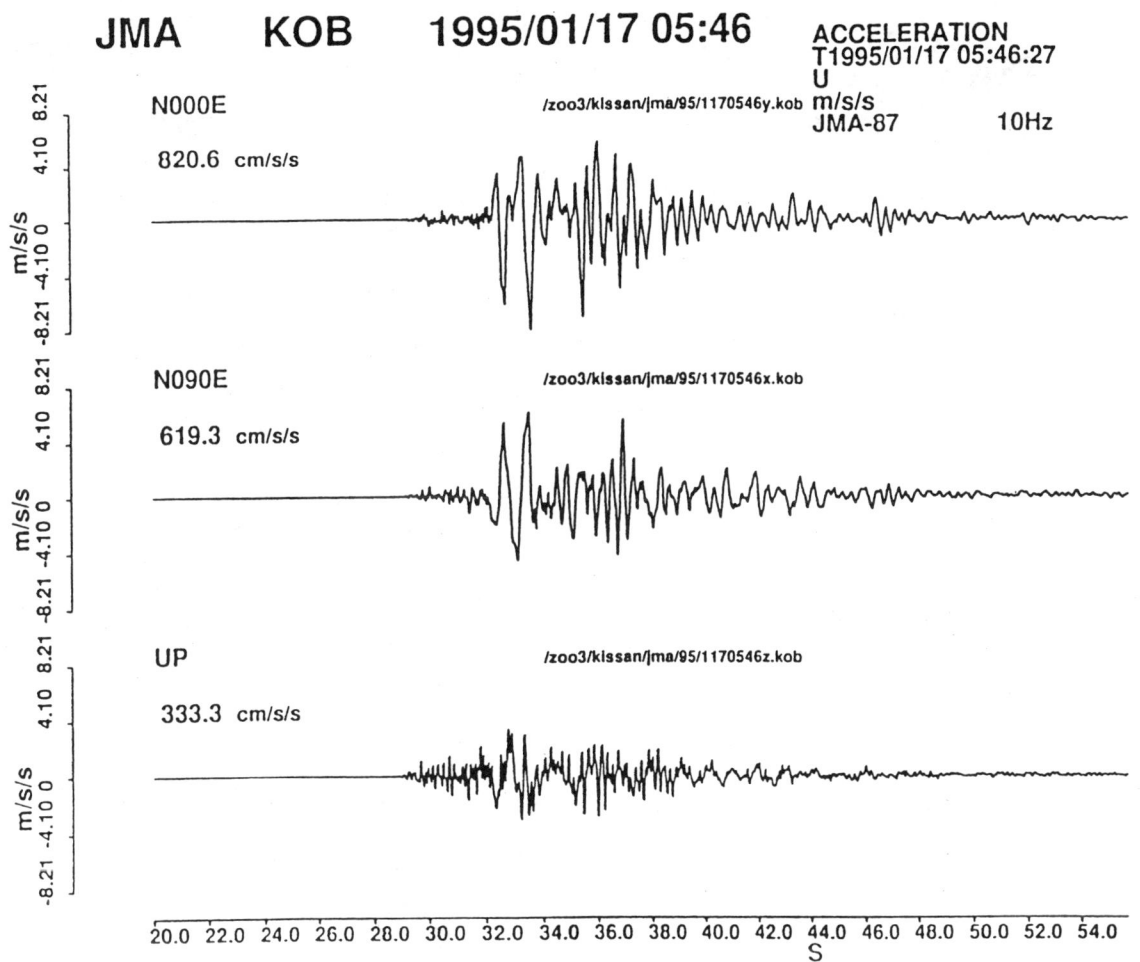

**FIGURE 2-4
GROUND MOTION RECORDED AT DOWNTOWN
KOBE ACCELEROMETER STATION
(DPRI, 1995b)**

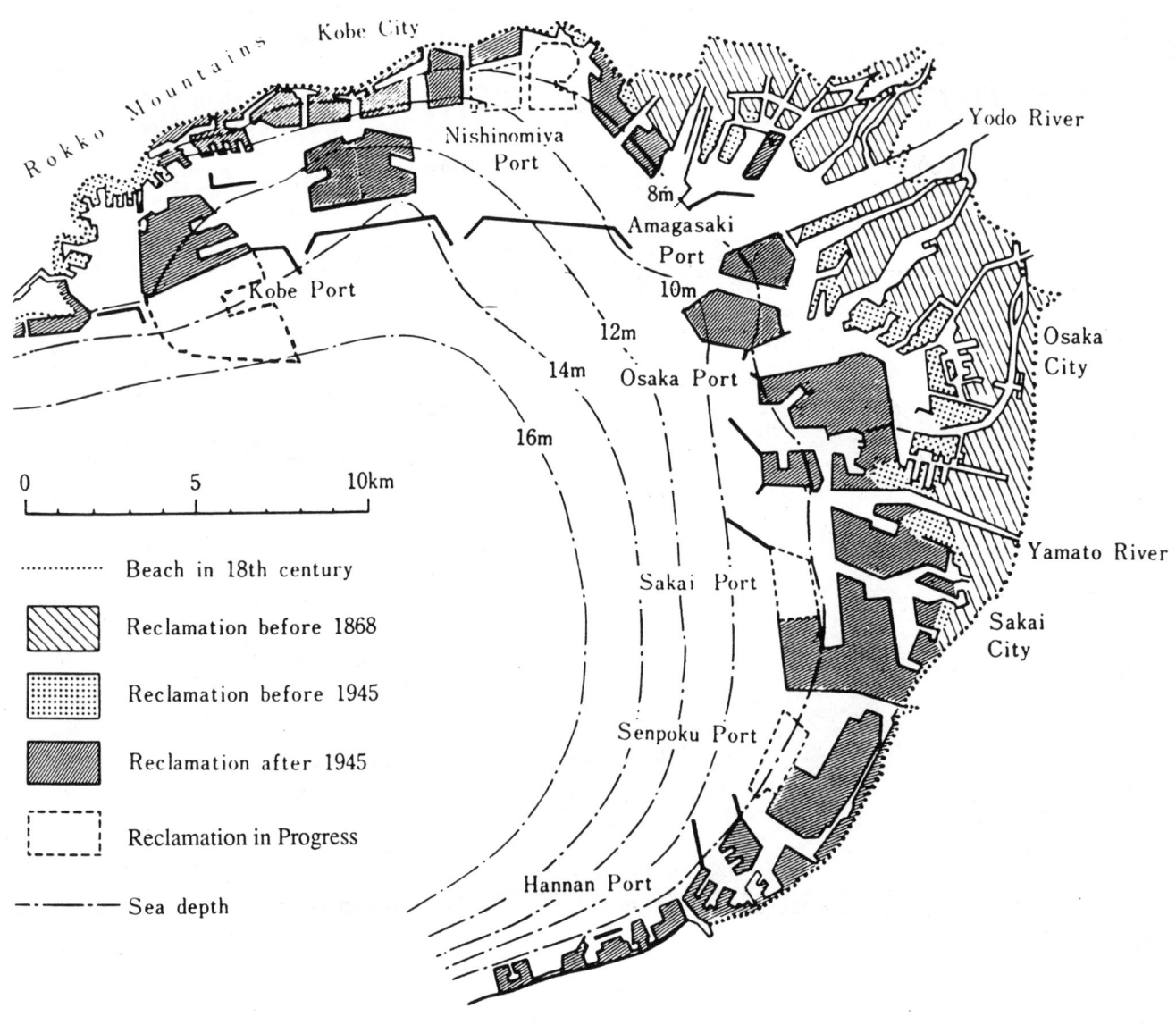

**FIGURE 2-5
RECLAMATION IN OSAKA BAY
(Mikasa and Ohnishi, 1981)**

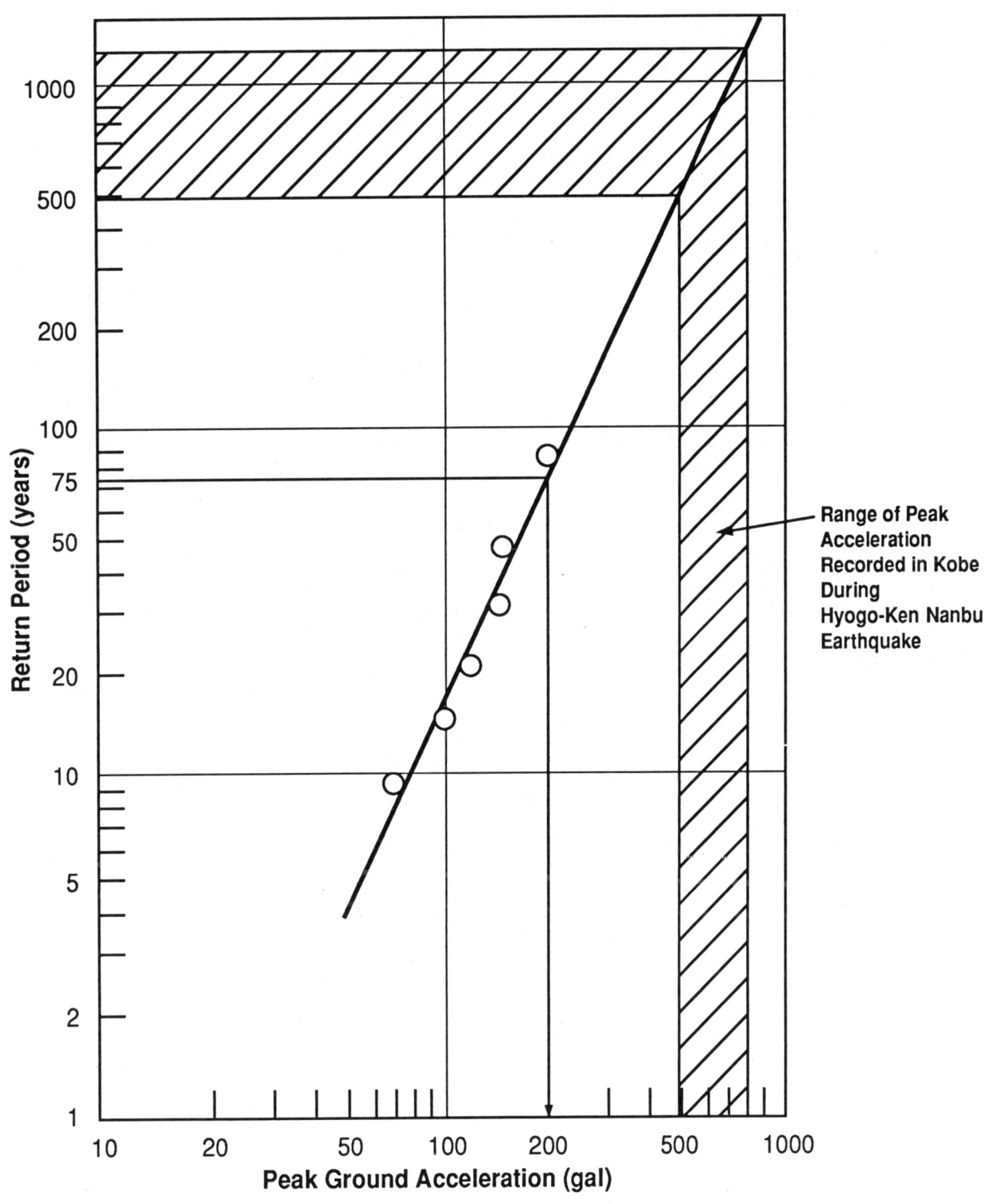

**FIGURE 2-6
PROBABILISTIC ESTIMATES OF PEAK GROUND ACCELERATION
IN JAPAN AND KOBE AREA (PHRI, 1995)**

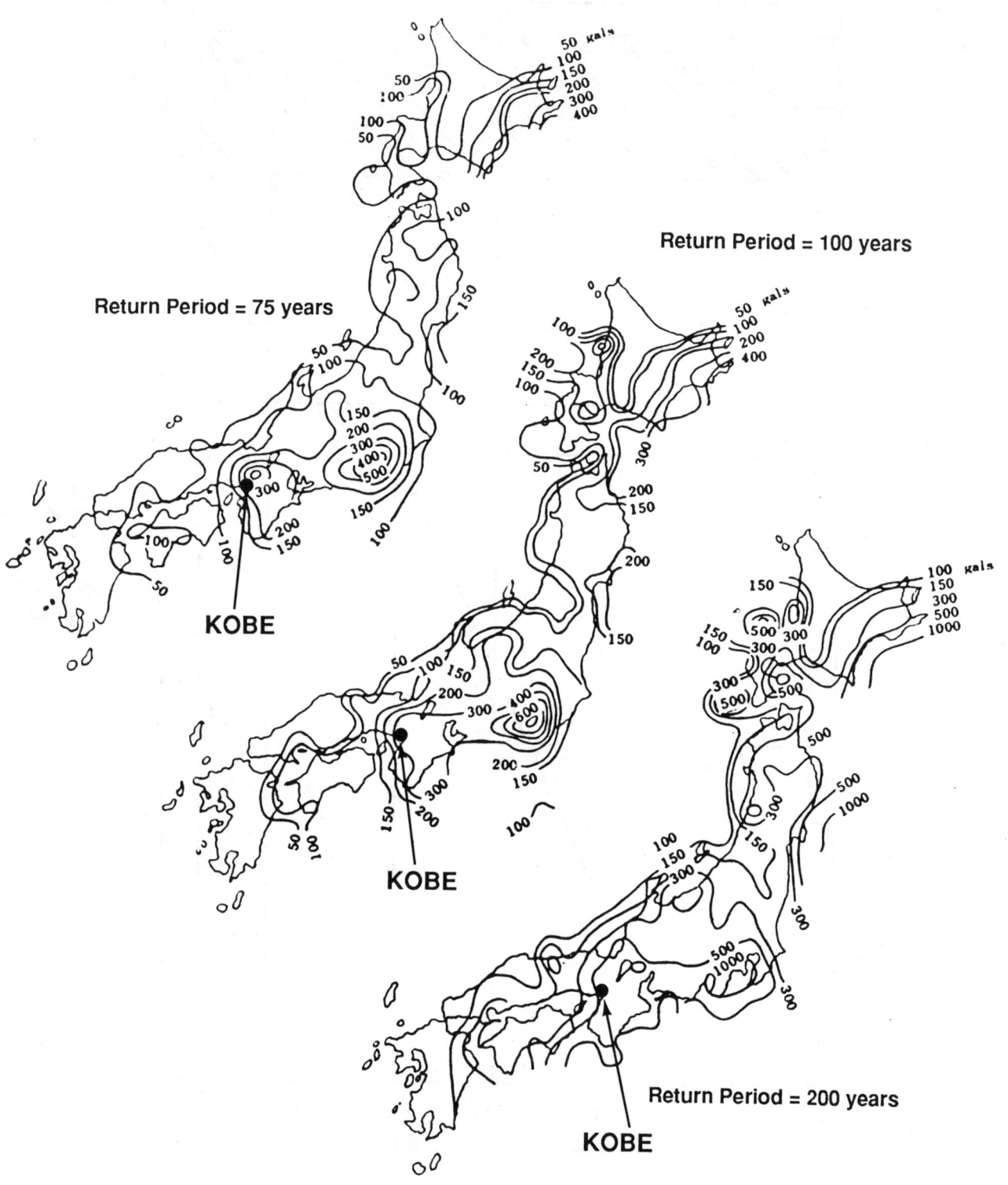

**FIGURE 2-7
PROBABILISTIC ESTIMATES OF GROUND SHAKING
THROUGHOUT JAPAN
(Kawasumi, 1951)**

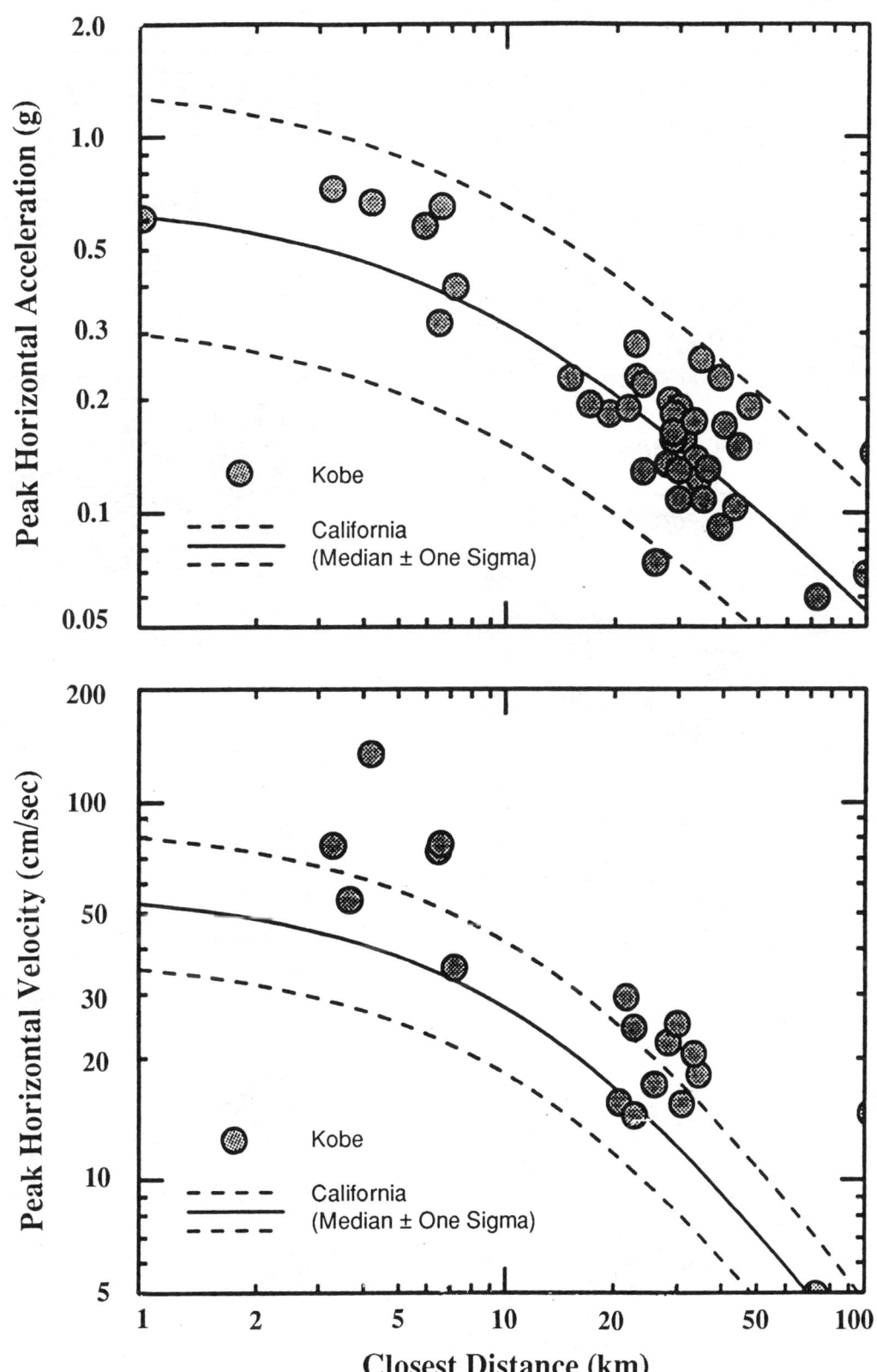

**FIGURE 2-8
COMPARISON OF PEAK ACCELERATIONS AND VELOCITIES
RECORDED IN KOBE TO THOSE PREDICTED USING EMPIRICAL
RELATIONSHIPS BASED MAINLY ON CALIFORNIA DATA
(SOMERVILLE, 1995)**

a) Building Collapse

b) Major Fires

c) Rail Bridge Collapse

e) Highway Bridge Collapse

d) Subway Collapse

**FIGURE 2-9
DAMAGE IN KOBE DUE TO HYOGO-KEN NANBU
EARTHQUAKE (INCEDE, 1995)**

# CHAPTER 3
# KOBE PORT

## 3.1 General Background

### 3.1.1 Port Description

The Kobe Port is Japan's largest container port, handling about 30 percent of Japan's container traffic, and currently ranks sixth worldwide in annual cargo throughput (53 million metric tons in 1993). The port contains approximately 64 km of total quay length, with about 9 km of this is devoted to container quay length and about 9 km for break bulk wharf and warehousing. As of April 1994, it had 24 container berths with facilities for accommodating up to 250 large ships at any time.

The Kobe Port is located along the northern margins of Osaka Bay. It extends in a predominantly east-west direction from the Eastern Sea Construction Areas 1 through 4 to the Suma and Nagata Harbors (West Reclaimed Lands) -- a shoreline distance of about 24 km (see Fig. 3-1).

With the exception of the western-most 4 km of the shoreline, the entire waterfront area of the port has been extensively developed for commercial use. The facilities that comprise the port include commercial zones (i.e., facilities managed and operated by the City of Kobe Port and Harbour Bureau), industrial zones (i.e., privately-owned facilities) and a relatively small marine zone (Meriken Park). The bulk of the cargo handling and storage facilities are located in six areas within the harbor limits, of which the largest are at Port Island and Rokko Island -- two major offshore reclamation islands (Fig. 3-2). The other four major cargo handling and storage facilities are located on the mainland and on near-shore reclamation islands, and consist of the Hyogo Piers, Shinko Piers, Maya Piers and Container Terminal, and Fourth Reclamation Area.

### 3.1.2 Chronology of Port Development

A chronology of the development of the Kobe Port is provided in Table 3-1, Figure 3-3, and the following paragraphs. It represents useful reference information for evaluating how such factors as design standards and construction methods may have affected the seismic performance of the port facilities during the Hyogo-Ken Nanbu Earthquake.

The port has been operating as an international port since 1868 and, since that time, development of new facilities and modernization of the older facilities has continued virtually uninterrupted. A period of extensive construction that started in 1897 and continued into the early 1940's led to the development of many of the port's mainland facilities including the Hyogo Piers, Takanawa Wharf and the Shinko Piers (No's. 1 through 6). The post-World War II era (from about 1950 to 1970) saw the completion of the Shinko Piers (No's. 7 and 8), the Maya Piers, and reclamation islands to the east that comprise the Eastern Sea Construction Areas 1 through 4.

Since 1970, the City and Port of Kobe have focussed development efforts on the two major offshore port complexes -- Rokko Island and Port Island. The first stage of Port Island was completed in 1981 after 15 years of construction, and Rokko Island was completed in 1991 after 20 years of reclamation work. Several redevelopment projects have also been carried out since 1970 to modernize selected existing facilities in other areas of the port.

With the completion of Rokko Island, development of the Kobe Port area has focussed on the planning and construction of a transportation system between various portions of the port. A 390 ha second stage of Port Island is currently under development, and the construction of an extensive system of bridges and tunnels that link the islands to the mainland has been initiated. In addition, plans are well underway for development of an approximately 300 ha island off of the southern shore of Port Island that will contain a new Kobe domestic airport.

### 3.1.3 Soil Conditions

The city of Kobe is founded largely on competent alluvial soils transported from the Rokko Mountains. These deposits are broadly characterized as an interbedded sequence of dense sands and gravels, and stiff clayey soils. The depth to bedrock beneath Kobe increases dramatically toward Osaka Bay to the south, due to dip-slip offsets of bedrock across numerous faults near the base of the Rokko Mountains. The portions of Kobe that have been built outboard of the historic margins of Osaka Bay are underlain by variable thicknesses of a soft to medium stiff marine clay which is very similar in its engineering properties to San Francisco Bay mud.

The Port of Kobe has been built almost exclusively on reclaimed land. Sandy soil has been placed over the soft clay deposits resulting in significant settlement due to the consolidation of clays. The thicknesses of the sandy fill and marine clay generally increase with distance south of the historic shoreline. The soil profile is somewhat similar to that found along the margins of the San Francisco Bay at the Ports of Oakland and San Francisco. A generalized soil profile for Port Island is shown in Figure 3-4, together with grain size distributions of sands throughout the Port of Kobe that liquefied after the Hyogo-Ken Nanbu Earthquake.

Port Island was developed by barge-dumping granular soil onto a roughly 10-to-15 m thick layer of very soft to soft marine clay. The fill material is predominantly decomposed granite (*masa* soil) that ranges in classification from SP/SM to SW/SM. Early reports on the performance of the reclamation at Port Island indicate that the fill includes an appreciable content of boulder-sized material (Nakakita and Watanabe, 1981). The fine-grain portion of the soil observed in numerous sand boils has negligible plasticity. The loose nature of the fill (as indicated by the low penetration resistance, $N_{avg} \approx 5$-$7$ blows/foot) resulted in an extremely high susceptibility to liquefaction, as will be discussed in a following section of the report. From the ground surface down, the soil profile at Port Island consists of approximately 15-to-20 m of loose sand underlain by 15 m of soft to medium stiff marine clay; 30-to-35 m of interlayered dense gravelly sand and stiff clay; 20 m of stiff marine clay; and interbedded very dense sand and stiff to hard clay to the maximum depth of the borings at 90 m.

Several methods of soil improvement have been utilized at selected sites on Port Island and Rokko Island. These techniques include preloading to minimize differential settlements under structures, sand drains, sand compaction piles and "composite" piles. The areas of the islands where soil improvement has been implemented are located primarily within the interior of the island (at areas of commercial development not associated with shipping and cargo handling). Figures 3-5 and 3-6 show that very little soil improvement has been performed at the shipping berths and wharves along the periphery of the islands.

The sand drains have been used primarily to expedite consolidation settlement of the marine clay that underlies the sandy fills. The sand compaction piles involve densification of the replacement sand, as well as, the soil adjacent to the piles. This soil improvement technique has been used to increase the bearing capacity and reduce the potential for earthquake-induced liquefaction of the sandy fill. Based on our visual inspection of several of the sites indicated as having received soil improvement, it appeared that the sand drains contributed negligible resistance to the development of excess pore pressures leading to liquefaction. This is not surprising in light of the placement method for the drains and the fact that densification of the soil is not provided by this method. Relatively few sites where sand compaction piles were used were accessible during our reconnaissance. Based on these limited observations, it appeared that the sites improved with sand compaction piles performed much better than adjacent sites which had not been improved. Further substantiation of the effectiveness of these soil improvement techniques for minimizing ground failures should await results of detailed instrumental surveys of the vertical and horizontal deformations at Port and Rokko Islands.

The review of a limited number of soil boring logs and profiles at other piers indicates that the soil profile at Port Island is fairly representative of the conditions at the other portions of the Port of Kobe if an allowance is made for the distance bayward from the historic shoreline as previously noted. For example, the Fourth Reclamation Area (island) is located at the eastern-most edge of the harbor limits of the Port of Kobe. The southwest corner of this island is approximately 2 km northeast of the Rokko Island Ferry Terminal. Boring logs obtained at a site located in the south-central portion of the island demonstrate that the soil profile consists of 4 m of medium-dense to dense (N = 20 to 55 blows/ft.) sandy (masa) fill underlain by approximately 13 m of loose to medium dense sandy fill ($N_{avg}$ = 8-to-10 blows/ft), 11 m of soft to medium stiff marine clay, and an interbedded sequence of medium dense to very dense sands and clayey sands to the depth of the boring at 42 m.

Although vibro-methods of soil improvement were reportedly used in soils adjacent to the concrete caissons along the southwestern portion of the island, it is not presently known if the upper 4 m of fill at the site of the soil boring (an inland location) was intentionally densified by a soil improvement technique or perhaps by a fortuitous construction sequence which densified the upper zone. The ground water table is located at a depth of 3 m; therefore it can be surmised that this near surface soil was probably end-dumped and compacted by construction traffic.

One potentially significant difference in the soil profiles was reported by several of the engineers that we met with. It was noted that the fill soil used at Port Island was sandy masa soil excavated from the Rokko Mountains northwest of Kobe, and that some of the fill used in the inland portions of Rokko Island contained "a significant portion" of crushed mudstone and siltstone material. The geotechnical properties of the crushed sedimentary rock fill used at Rokko Island have not yet been ascertained (as of April 1995). It is interesting to note that the grain size characteristics of soils excavated from sand boils located along the perimeters of Port and Rokko islands are very similar.

Based on the geotechnical information available at this time, it appears that most of the areas within the Port of Kobe are founded on similar soils. Among the soil characteristics that these areas have in common are: (a) thick surficial layers of loose saturated sand; (b) fairly extensive deposits of soft to medium stiff marine clays; and (c) deep soil profiles over basement rock. The extensive occurrence of liquefaction in the sandy fill at almost every pier within the Port of Kobe has been well documented. In addition to this phenomena, the underlying soils also had a significant influence on the characteristics of the strong ground motions experienced throughout the port. Although very few acceleration response spectra are available at this time, it is anticipated that the deep soil deposits contributed to an enhanced intensity of strong motion at longer periods (on the order of 0.5 sec. to 1.0 sec or greater) and a reduced intensity of shaking at the shorter periods. These amplified longer period motions probably contributed to the damage that was observed at most of the bridges which connect the large reclamation islands.

### 3.1.4 Caisson Quay Wall Sections

Typical caisson quay wall cross-sections along Rokko Island and Port Island are shown in Figure 3-7. This figure shows that these quay walls consist of concrete caissons that are filled with sand and supported on a cobblestone foundation that rests on the underlying sandy fills. The backfill that has been placed along the landward side of the walls consists of gravelly material adjacent to the wall and sandy soil further inboard. Pile supports have not been provided for any of the quay walls at Port or Rokko islands. Piles were used only for the landward crane rail at the crane locations along the west and north faces of Port Island. None of the crane rails at Rokko Island are pile supported.

The quay walls along the smaller islands and the mainland also consist of concrete caissons whose cells are filled with sand and, in some cases, concrete. Unlike Port Island and Rokko Island, pile foundations have been utilized to support several sections of the mainland quay walls. Examples of these foundation types include the Naka Pier and Pier 6 of the Shinko Piers (Figure 3-8). Piles are also used to support piers at the Takahama Wharf and at Berth C, Pier 1 of the Maya Piers, as discussed in a subsequent section of this report. Steel pipe piles have apparently been used to support crane rails along the eastern portions of the Maya Piers (Berths Q, R and S).

### 3.1.5 Seismic Design Procedures for Caissons

The seismic design procedures used for caissons at Japanese ports have been based primarily on an evaluation of the stability of the walls when they are subjected to inertia forces that are defined in terms of an equivalent seismic coefficient (i.e; psuedo-static lateral force methods). Code-based methods of analysis did not address the effects of liquefaction of foundation and backfill soils on the seismic performance of the caissons and other waterfront retaining structures (i.e.; quay walls, piers, breakwaters, bulkheads) until the most recent revision of the design standards was adopted in 1989 (JPHA, 1989). To our knowledge, such assessments had not been incorporated into the prior design of any of the caissons within the Port of Kobe.

*(a) Equivalent Seismic Coefficient*

The caisson-type quay walls at the Port of Kobe have been designed to resist psuedo-static lateral forces computed using a variety of seismic coefficients. For example, seismic coefficients used in the design of the quay walls at Port Island and Rokko Island were 0.1 and 0.15, respectively. At the Maya Pier, most of the quay walls were designed using a seismic coefficient of 0.18, except along the west side of Pier No.1, where a seismic coefficient of 0.25 was used. Caissons at the southwest portion of the Fourth Reclamation Area were designed with a seismic coefficient of 0.15. Based on the information provided to us in Japan, it appears that the remaining quay walls along the mainland piers and wharves were designed using a seismic coefficient of 0.1.

In order to evaluate the design criteria utilized for waterfront retaining structures at various portions of the Port of Kobe, a comparison of the seismic design codes prescribed for port and harbor facilities in 1959 and 1978 is presented. The zonation maps which provide the seismic coefficients for Japan are illustrated in Figure 3-9. In both maps, Kobe and Osaka lie in the region rating the highest seismic coefficient. It is interesting to note that the range of seismic coefficients was reduced from 0.15-0.25 to a single value of 0.15 in the latter code. The sparse documentation discovered to date on the 1959 code (OCSWCEE, 1960) indicates that;

> *"The ranges of the seismic coefficient in each section (i.e., seismic zone) are legislated, and the final seismic coefficient used in the design of a structure is decided from the standard range of seismic coefficient, taking into consideration the kind and importance of the structure, and the condition of the foundation."*

No information was available regarding the factors used in the code to represent the soil conditions and the importance of the structure.

The most recent code that is currently available (1989) for port facilities defines the design seismic coefficient ($k_h$) as:

$$k_h = ZGI$$

where Z is the seismic zone factor, G is the ground condition factor, and I is the importance factor. The Z factors used in this computation of $k_h$ are shown in Figure 3-9b, and the G factors and I factors are given in Table 3-2. The factors of safety incorporated into the seismic design of the caissons are 1.0 and 1.1-to-1.2 for sliding and overturning respectively.

As of this writing, information describing the basis for specifying the seismic coefficients in areas such as the Maya Piers, where several different values have been used for the design of the quay walls and the pile supported berth has not been available. It is anticipated that such issues will be resolved in the near future.

*(b) Assessment of Liquefaction Potential*

The most recent (1989) Japanese code for port facilities now includes provisions for assessment of the liquefaction potential of saturated sandy soils during earthquakes. These provisions consist of the following steps (Tsuchida, 1990); (a) check the grain size distribution of the soil against critical ranges of grain sizes specified by Iai et. at. (1989) for liquefiable soils (Figure 3-10a); (b) compute equivalent penetration resistances (N-values) for each saturated sand layer by correcting N-values from SPT test results for fines content and to a reference vertical effective stress (0.66 $kgf/cm^2$); (c) carry out an equivalent linear site response analysis, and obtain an effective acceleration for each saturated sand layer as described by Tsuchida; (d) for each layer, use the corrected N-value and the computed effective acceleration to enter a design chart (Figure 3-10b) in order to assess the potential for liquefaction of the soil at the site; and (e) if liquefaction is indicated, incorporate appropriate countermeasures to reduce liquefaction risks to the port. Because these liquefaction assessment procedures have been in effect in the codes for ports since only 1989, it is our understanding that site-specific liquefaction analyses have been carried out for only a very few locations (on Port Island only) prior to the Hyogo-Ken Nanbu Earthquake (Youd, 1995).

## 3.2 Earthquake Ground Motions

The general level of ground shaking at the Port of Kobe is indicated by the ground motions recorded at the Kobe Port Construction Office (on the mainland) and by a downhole array of accelerometers on Port Island. The accelerograms at the Kobe Port Construction Office (Figure 3-11) exhibit peak horizontal accelerations of 502 $cm/sec^2$ and 205 $cm/sec^2$, and a peak vertical acceleration of 283 $cm/sec^2$. The duration of the strong shaking segment of the recorded horizontal motions was about 5 sec.

The downhole array of accelerometers on Port Island are located at the northwest portion of the island, just south of the Kobe Bridge (Figure 3-12a). The array consists of instruments on the ground surface and at depths of about 17 m, 33 m, and 84 m; the uppermost two instruments are at the top and the base respectively of the loose fills, the third instrument is in an intermediate sand layer that is medium dense, and the deepest instrument is in a very stiff sand layer (Figure 3-12b). The horizontal motions recorded at each instrument in this array (Figure 3-13) were very strong, with peak accelerations that ranged from about 280-340 $cm/sec^2$ at the ground surface and larger values at various depths below the ground surface (about 540-680 $cm/sec^2$). Vertical motions were also recorded at this array and had peak accelerations of 560 $cm/sec^2$ at the ground surface, 790 $cm/sec^2$ at the 17 m depth, and about 190 to 200 $cm/sec^2$ at lower depths. The accelerations at the top of the fill were smaller and had a much longer predominant period than the accelerations at the base of the fill. This reflects the effects of softening, and liquefaction of the upper fill layer.

Based on the pattern of peak accelerations throughout the city of Kobe, the relative distances to the rupture zone, and the similarities in the soil profiles, it is inferred that the range of strong ground motions were similar across Port and Rokko Islands.

## 3.3  Seismic Performance Overview

The Port of Kobe was severely damaged during the Hyogo-Ken Nanbu Earthquake. Virtually all of the 240 berths at the port were closed indefinitely after the earthquake (i.e., only 6 of the 240 berths were serviceable to any degree). Repair costs for the port have been estimated at about $10 billion, and repair times (to restore total operations at the port) have been estimated to be about 2 years from the time of the earthquake (Griffin, 1995). Probable causes of this severe damage and the types of damage that occurred at the Kobe Port are discussed in the remainder of this chapter.

## 3.4  Seismic Performance of Waterfront Retaining Structures and Cranes

The primary cause of the extensive damage at the port was ground failure due to widespread liquefaction of the fill materials (Fig. 3-14). This resulted in increased lateral pressures applied to the quay walls which, for the quay walls without pile supports, led to large seaward displacements of the walls (often on the order of several meters). As a result of these seaward movements, the fill soils behind the quay walls moved laterally and settled substantially (Fig. 3-15). The large lateral soil deformations in the waterfront areas extended into the backland areas distances of up to 75-to-100 m, as evidenced by tension cracks in A.C. pavements and movement of near surface soils away from pile-supported structures. Youd (1995) has reported that ground fissures oriented parallel to the quay walls formed as far inland as 200 m, indicating the extent of the lateral spreading.

### 3.4.1  Concrete Caisson Quay Walls

Most of the quay walls at the Kobe Port were concrete caissons (soil and concrete filled) without pile supports, as described in Section 3.1.4. The seismic performance of these types of quay walls in the presence of the liquefied fills and increased lateral pressures against the walls was poor. In addition, liquefaction of the underlying foundation soils may have contributed to the poor performance of these quay walls.

Based on information obtained to date, we are aware of only two locations where concrete caisson quay walls were supported on piles -- the southern end of the Naka Pier and along the shallow-water sections of Shinko Pier No. 6. At each of these locations, foundation support is provided by tapered timber piles with a tip diameter of 0.21 m and lengths ranging from 6.3 m to 7.2 m. The ground deformations at these pile-supported caissons were slightly-to-significantly less than those observed at adjacent caisson quay walls without pile supports. It is noted that similar piles were placed beneath the concrete caissons at Hyogo Pier No. 2 although, based on the cross sections provided to us during our reconnaissance, it does not appear that the piles actually support the caissons at this pier. This is significant because the movement of the caissons at Hyogo Pier No. 2 was large, resulting in complete submergence of the end of the pier.

Regarding the effects of the seismic design coefficient ($k_h$) on the performance of these caisson quay walls, we were informed that the $k_h$ values used in the design of the caissons at Port Island and Rokko Island were 0.1 and 0.15, respectively. These different $k_h$ values are partially reflected by the smaller height-to-width (H:W) ratios for the caissons at Rokko Island. (The values of these ratios ranged from 1.21 to 1.32 at Port Island and from 0.97 to 1.19 at Rokko Island.) It is important to note that no form of soil improvement explicitly intended to densify the foundation pad or backfill soils adjacent to the caissons was performed at either Port or Rokko Islands. For the most part, the performance of the caissons during this earthquake complements an extensive list of case histories that demonstrate the deficiencies of seismic coefficient-based design methods for caissons in loose saturated sandy soils, since these methods do not account for the potential liquefaction of these soil materials (Werner and Hung, 1982).

### 3.4.2 Cranes

All of the gantry cranes at the Port of Kobe are rail mounted. Most of these cranes have a rail span of 30.5 m, and were built in the 1980s and 1990s. A few of the older cranes at Port Island (built in the early 1970s) have a rail span of 16 m. All bayward crane rails rest on the caisson. The landward rail for the cranes with a 30.5 m rail span were supported on engineered fill, and the landward rail for the cranes with a 16 m rail span were supported on piles. Most of the cranes were secured with the stowage pins in place (Liftech, 1995).

The quay wall displacements (translational and rotational) and the large differential soil movements described previously resulted in severe spreading and deformation of crane rails. This, in turn, led to buckling and yielding of the legs of many of the cranes, and complete collapse of the cranes in some cases (Figures 3-16). Although liquefaction and movement of the underlying soils was the primary cause of the observed crane damage, Liftech (1995) has observed that the gantry crane construction also influenced the degree and type of damage experienced by the cranes during the Hyogo-Ken Nanbu Earthquake. In particular, it was noted that cranes with weak portal beams and with non-ductile moment frames suffered significant buckling of the legs above the portal frames, whereas damage to cranes with strong portal beams forming ductile moment frames was restricted to small areas near the portal beams.

### 3.4.3 Steel Plate Cellular Bulkheads

Instead of the concrete caissons which are predominant throughout the port, steel plate cellular bulkheads were deployed at the Maya Piers during their original construction in the late 1950s and early 1960s. These caissons were fabricated out of 9 mm thick steel plate, and were formed to a diameter of 15.5 m and height of 14.0-to-16.5 m. The cells were filled with sandy soil and 0.4-to-0.5 m diameter steel pipe piles were driven through the interior fill to support crane rails. It is noted that, during the subsequent redevelopment of the Maya facilities for container handling and storage (1987-1991), concrete caissons were placed outboard of many of these older cells.

Modes of failure of the steel cell caissons are similar to those outlined for concrete caissons (i.e., settlement due to densification and deformation of foundation soils, translation caused by increased earth pressures, and loss of ground behind the retaining structures due to the flow of liquefied soil through gaps between individual caissons). This is shown in Figure 3-17.

### 3.4.4 Concrete Block Quay Walls

In many of the shallow-water portions (i.e., inboard ends) of the piers at the Port of Kobe, retaining walls constructed out of stacked blocks of concrete are used. The blocks are generally 1.5 m in height and increase in width with each successive lower block (e.g.; at the Maya Piers, the width of the upper block is 1.5 m and the lower blocks measure 2.5, 3.5, 4.5, and 5.0 m; at the Naka and Shinko Piers, the blocks measure 1.32, 1.52, 1.82, and 2.93 m). The performance of these retaining structures is also linked to the extent of the liquefaction that occurred during the earthquake. We observed damage to these walls that varied from minimal to catastrophic, with greater damage experienced at the Naka and Shinko Piers.

### 3.4.5 Quay Walls at Maya Piers

The development of the Maya Piers, located near the mainland between Port and Rokko Islands (Fig. 3-18), was initiated with reclamation of the island in 1959. Construction was completed in 1967, and the facility originally consisted of four finger piers extending to the south (i.e.; bayward) with additional shallow-water berths (4 m) along the northern perimeter of the island. A variety of waterfront retaining structures have been employed at the Maya Pier complex, providing an opportunity to evaluate the relative seismic performance of these structures under equivalent seismic loads.

An extensive collection of quay wall sections for the Kobe Port which we obtained during our reconnaissance (Iwasaki, 1995a; Tsuchida, 1995), indicates that the waterfront areas of the Maya Piers were originally developed as follows. In this, it is noted that the locations of the various berths denoted below are shown in Figure 3-18.

(a) At Berths D, G, I, K, and Q, steel plate cellular bulkheads (SPCBs), with a diameter of 15.5 m and a height of 16.5 m were deployed.

(b) Along the south-facing quay of Pier No. 2 and at Berth O, SPCBs were deployed that had a diameter of 15.5 m and a height of 14 m.

(c) At Berth A, SPCBs with a diameter of 15.5 m and a height of 15 m were deployed, together with an outboard wharf supported on batter piles.

(d) Along the south facing quay of Pier No. 1, soil-filled concrete caissons were deployed that had a width of 9.1 m and a height of 10.7 m.

(e) Along many of the shallow-water (4 m) portions of the piers, concrete block walls were deployed.

(f) SPCBs were also used along Berths B, C, E, H, J, R and S, although information on the exact dimensions of these cells was not available at the time this report was prepared.

In all cases, the SPCBs were founded on rubble (i.e., rock) fill and backfilled with sandy soil. Steel pipe piles (with a diameter of 40-to-50 cm, a wall thickness of 0.6 cm, and a length of 24-to-30 m) were driven through the interior fill to support crane rails. In most cases, a row of interlocking steel pipe piles (with a diameter of 40 cm, a wall thickness of 0.9 cm, and a length of 30 m) were driven adjacent to the bayward portion of the cells to form a continuous seawall.

In 1987, a four year redevelopment project was initiated to transform Piers No. 3 and 4 into a container handling facility - the Maya Container Terminal. At that time, the 9.5 ha area between the two piers was filled, and new concrete caissons were placed roughly 15 m bayward of the older SPCB wall at Berths O and P. The space between the caisson and the SPCB was filled with sandy soil. The soil was presumably end dumped, and it was not densified prior to the construction of the pavement section for the deck.

The facilities along the western side of Pier No. 1 were also expanded during this recent phase of development. At Berth A, the pile-supported wharf has been removed and replaced with 17.1 m by 15 m concrete, soil-filled caissons (Fig. 3-19a). The caissons were placed approximately 3 m outboard of the older SPCB on an asphalt mat underlain by rubble fill. The void between the retaining structures was filled with rubble. A similar procedure was followed at Berth B, where 14.4 m by 12 m concrete, soil-filled caissons were placed bayward of the original SPCB. Rubble fill was again used to fill the area between the retaining structures.

At Berth C, a pile-supported wharf was constructed in lieu of the concrete caissons used at Berths A and B (Fig. 3-19b). A 10 m wide wharf supported by 1.2 m steel pipe piles was constructed adjacent to the original SPCB. The inboard row of these piles includes batter piles, which have often been identified as particularly vulnerable to strong shaking during earthquakes (Werner and Hung, 1982; EERI, 1990). The outer row of piles consists of interlocking steel pipe piles which form a continuous wall (Figures 3-20k and l). The area between the interlocking piles and the original SPCB has been filled with rubble to a level that was below the tops of the piles (see Fig. 3-19b).

Based on the information currently available, it appears that the area between Piers No. 1 and 2, Berths D and E, and the southern edge of Pier No. 1 were not modified during redevelopment. In these areas the original quay walls, designed with static lateral force coefficients of 0.18, are still in place.

The new caissons and pile-supported wharf at Berths A, B, and C of Pier No. 1 have been designed using a seismic coefficient of 0.25. The use of this relatively high static lateral force factor has led local engineers to refer to this section of the Maya Piers as the "Aseismic-Reinforced Berths" or the "Earthquake Resistant Berths". In striking contrast to the poor performance of the other waterfront areas at the Maya Pier complex (as summarized below), the quay walls at Berths A, B and C exhibited very little damage during the earthquake.

The photographs in Figure 3-20 provide an overview of the damage sustained by the various quay walls and pile-supported structures at the Maya Piers. With the exception of Berths A, B, and C, damage to quay walls was extensive throughout the Piers. At the north approach to the Piers, lateral movement of the quay walls resulted in damage to adjacent building and warehouse foundations (Figures 3-15f, 3-20a, b). Liquefaction-induced damage to the waterfront structures was also extensive along the northern perimeter of the island. The new concrete caissons along Berths O and P shifted dramatically toward the bay, resulting in minor warping of the legs of cranes working in this area (Fig. 3-20c). To the west, similar damage to concrete block retaining structures between Piers No. 1 and 2 was experienced (Fig. 3-20d).

Pier No. 1 provides a valuable opportunity to evaluate the effectiveness of various types of retaining structures and seismic design provisions for waterfront structures. The waterfront areas along the east and south-facing portions of the pier were not affected by the redevelopment. It is again noted that the seismic coefficients used in design of the original caissons and the newer retaining structures are 0.18 and 0.25, respectively. The quay wall at Berth E failed dramatically with the retaining structure translating and rotating into the bay. The adjacent SPCB at Berth D performed relatively well, although the backland soils liquefied and appear to have flowed into the bay through gaps between the cells. (Figs. 3-17a, b). The southern corner of the pier includes a pile supported wharf. Lateral ground deformations resulted in buckling of several of the steel pipe piles (with a diameter of 1.0 to 1.2 m) near the pile caps. The concrete caissons along the southern end of the pier moved toward the bay; this resulted in the formation of a deep graben (2-to-3 m deep) in the backland area, but surprisingly little damage to the adjacent warehouses (Figs. 3-20f, g). It is not until the quay wall at Berth A is approached that any appreciable difference in the seismic performance of the retaining structures is apparent.

As previously noted, the quay walls along the west side of Pier No. 1 have been reconfigured with the addition of concrete caissons or interlocking steel pipe piles. These retaining structures are founded on thin mats ($\approx 0.5$ m) of rubble fill and dredge sand. The retaining structures at Berths A, B and C performed very well during the earthquake, with only minor deformations observed between Berths B and C (Figs. 3-20h-j). No damage of the interlocking pipe pile wall at Berth C was evident (Figs. 3-20k, l). The continuous nature of the waterfront wall at Berth C precluded our observation of the batter piles which support the inboard portion of the wharf.

The seismic resistance of the quay walls at Berths A, B and C provides one of the few examples of acceptable performance of waterfront retaining structures at the Kobe Port. The use of stringent psuedostatic design requirements is considered to be only partially responsible for this success. Liquefaction of the fill surrounding the warehouses adjacent to the quay walls was evidenced by sand boils and substantial settlement. The increased lateral earth pressures due to liquefaction of the fill were resisted by both the SPCB and the newer retaining structures. It is surmised that the rubble fill immediately behind and beneath the quay walls was not susceptible to the generation of significant excess pore pressures. Therefore, the two retaining structures, acting together, provided adequate resistance to the increased lateral earth pressures.

An interesting comparison can be made between the performance of the new quay walls at Berths A, B and C and the new caissons at Berth O and P, where concrete caissons have also been placed outboard of the original SPCB. Based on information obtained during our reconnaissance, the seismic coefficient used for the new caissons at Berths O and P was 0.18. This may have contributed to the disparate performance of the caissons at this portion of the Maya Container Terminal. More important is the fact that the new concrete caissons at Berths O and P were backfilled with loose sand, not the coarse rubble used at Berths A, B and C. Liquefaction of the fill behind the caissons at Berths O and P clearly contributed to the failures that occurred.

## 3.5. Seismic Performance of Pile-Supported Structures

### 3.5.1 Pile-Supported Structures in Waterfront Areas

Based on available information at the time of this report preparation, nearly all of the port's wharves, warehouses, passenger ferry terminals, commercial buildings, and elevated highway and rail systems are supported by end-bearing piles. These piles extend through the loose sandy fill and underlying soft marine clay to depths of 20-to-30 m or greater where they are embedded in the dense older alluvium. The placement of the fill over the marine clay has resulted in considerable settlement due to consolidation of the clay (e.g.; 4 to 5 m at Port Island). Most structures are supported by deep foundations in order to alleviate potential damage due to differential settlements.

The extensive catalog of quay wall sections and a very limited number of foundation schematics for bridges, buildings and other structures, indicate that virtually all of the piles are vertically oriented. The only use of batter piles that has been documented to date is at Berth C, Pier 1 at the Maya Piers. These batter piles are isolated from view by a row of interlocking steel pipe piles; therefore, the condition of these piles and the pile cap had not been ascertained at the time of our reconnaissance.

The observed performance of piles and pile-supported structures can be related in most cases to the direction and extent of ground deformation which occurred at the site. As previously discussed, the soils adjacent to quay walls and extending significant distances inland exhibited pronounced lateral as well as vertical deformation. In these areas, piles were often subjected to large lateral loads due to the movement of the surrounding soils. Soil settlements facilitated our observation of numerous piles and pile-caps beneath structures located in waterfront areas. Common concrete pile types included roughly 30 and 40 cm diameter hollow cylinder piles (approx. 8 cm wall thickness), 30 and 40 cm diameter solid cylinder piles, and 55-to-70 cm diameter hollow cylinder piles (approx. 8 cm wall thickness). In most cases, steel reinforcement for both prestressed and conventionally reinforced concrete piles was minimal or nonexistent.

During our reconnaissance, we observed numerous examples of severe cracking or fracture of hollow concrete cylinder piles at or near their connection to the pile cap (Fig. 3-21b). In addition, we observed a number of solid concrete cylinder piles in the waterfront area that appeared to perform well in the presence of large lateral movements of the surrounding soils, with only negligible-to-minor hairline cracking near pile caps. This contrasting performance of the hollow concrete cylinder piles and the solid concrete cylinder piles has also been documented by members of other reconnaissance teams (e.g., Youd, 1995).

Steel pipe piles were also observed to have experienced damage. Underwater inspection of a pile-supported grain transport wharf at the southwestern margin of the Fourth Reclamation Area revealed moderate buckling to 1.2 m diameter (1.9 cm wall thickness) steel pipe piles (Fig. 3-21a). Damage at this wharf included the collapse of cargo cranes (presumably due to inertial effects as opposed to the ground failure-induced mode of failure common at Port and Rokko Islands), minor deformation of crane rails, and severe damage to the grain conveyor system. Damage to the wharf deck itself appeared to be minor. An example of apparently good performance of steel pipe piles is the vertical piles that support the Takahama Wharf, whose deck did not suffer any cracking or damage (Fig. 3-22).

During our reconnaissance, we observed that the lateral and vertical resistance provided by the piles invariably minimized the lateral and vertical displacements and deformations of waterfront buildings (e.g., Fig. 3-23c). Although damage to some of the pile-supported buildings due to ground shaking was observed, there was relatively small permanent displacement of the building structures even though the surrounding soils moved substantially. In waterfront areas, the lateral ground deformations decreased with increasing distance from the quay walls. Soils underlying waterfront structures subsided and settled substantially throughout the port facilities. In several cases, the pile foundations for large warehouses exhibited extensive damage on the waterfront side of the structure (Fig. 3-21b) and minimal damage on the inland side. In many instances, the warehouses performed very well despite the extensive deformation of foundation soils and complete separation from the pile foundations. This performance may also be attributed to the existence of well-reinforced floor slabs.

The influence of pile foundations on the extent of lateral movements was also observed at several large bridge foundations. Lateral ground deformations were extensive (greater than 1-to-3 m) adjacent to the foundations of many of the major bridges which link the port facilities. In every case, the extensive pile groups limited the deformation of the bridge piers (as discussed further in Section 3.7.2).

It should be noted that our observations regarding waterfront pile performance (particularly where the piles appeared to be undamaged) should be tempered by the fact that our observations were confined to the upper several feet of the piles, and that only very limited underwater or subsurface inspections of piles had been made at the time of our reconnaissance. In view of the large soil movements that occurred along the waterfront area of the Kobe Port, it is likely that many more examples of pile damage could be uncovered after more extensive subsurface inspections of the piles are carried out. The performance of the piles within the liquefied soils, at the interface between the sandy fill and the soft marine clay, and at the interface between the marine clay and the older alluvium, are important issues that remain to be addressed. Assessment of these particular issues, together with the overall assessment of the seismic performance of the waterfront pile foundations, should also include the compilation of further information pertaining to the type of piles employed, the geometry of the pile groups, and the design and construction of the piles.

### 3.5.2 Pile-Supported Structures in Interior Areas

Ground deformations in the inland portions of the reclamation islands were predominantly vertical due to the densification of the sandy fill. In these areas, pile-supported structures appeared to perform very well, and their foundations did not exhibit any evidence of permanent deformations that could be related to pile failure. On several of the reclamation islands, pile-supported buildings appeared to remain at their design elevations despite significant settlement of surrounding soils. On Port Island, the relative vertical movement averaged roughly 0.5 m, with settlements in several portions of the island reaching as much as 1 m (Fig. 3-23a, b, and d). This relative movement between the pile caps and the ground surface is interpreted to indicate that the piles did not fail by buckling in the liquefied soils, nor did they settle appreciably in the dense bearing strata due to the loss of skin friction in the liquefied fill and subsequent down drag on the pile that may have occurred as the densified fill settled. The latter phenomenon is judged to be a minor effect, since excess pore pressures are still high as the sand settles and it is anticipated that gaps may form between the soil and the pile in the upper several diameters during strong ground shaking. Both of these factors would significantly reduce the downdrag due to skin friction on the pile immediately after the earthquake. From a practical perspective, an additional factor that would contribute to minimal earthquake-induced settlements of end bearing piles (which is the predominant pile type at the Kobe Port) is the relatively large factor of safety that is commonly used in the design of deep foundations. It is also common for geotechnical engineers to ignore the contribution of skin friction provided by loose soils, such as the sandy fill, to the load carrying capacity of the end bearing piles. These design methods would also result in a substantial reserve in the end bearing that could be provided by the dense clayey sands at depth. It is again noted that, at the time of our reconnaissance, extensive subsurface inspections of pile foundations had not been made. Such investigations may reveal pile damage that was not evident during our reconnaissance.

As a related issue, the settlements observed next to pile-supported structures provide a unique opportunity to evaluate current methods for estimating settlements of sandy soils due to earthquake shaking (e.g.; Tokimatsu and Seed, 1987; Ishihara and Yoshimine, 1992). Given the penetration resistance of the sandy fill at two sites on Port Island (Figs. 3-4 and 3-12 show an average value of $(N_1)_{60}$ = 7 blows/ft) and the accelerations recorded in the downhole array (Fig. 3-13), the Tokimatsu-Seed method predicts volumetric strains of the sand of about 3%, while the Ishihara-Yoshimine approach indicates a post-earthquake volume reduction of about 4%. For a fill thickness of 17 m, (as shown in Fig. 3-12), these strains correspond to ground settlements of 0.5-to-0.7 m. These estimated values agree well with observed settlements, indicating the utility of these methods for estimating earthquake-induced settlements in sandy soils.

In the inland areas of Port and Rokko Islands, newer multistory buildings (most likely designed to post-1980 structural codes) appeared to perform very well during the earthquake. Judging from the exterior appearance of the structures, architectural damage was minor and no evidence of permanent deformation was observed. In light of the substantial ground settlement which occurred next to these buildings, it is anticipated that underground utility lines and other appurtenances could have been damaged. We understand that the extent of such damage to structures and buried lifelines throughout the port complex is under continued investigation.

## 3.6  Seismic Performance of Buildings

### 3.6.1  Older Corrugated Metal, Wood, and Non-Ductile Concrete Frame Buildings

Older low-rise buildings (e.g., warehouses, etc.) within the port were subjected to varying degrees of damage due to the earthquake-induced inertia forces that greatly exceeded those considered during the design of these structures. The port buildings that fall in this category were constructed of corrugated metal, wood, or concrete shear wall or moment frame (probably non-ductile) elements. Damage to the corrugated metal and wood buildings took the form of connection failures, large racking deformation of the walls (which caused jamming of doors into the buildings), foundation damage due to large ground movement, and falling of contents of the buildings (Fig. 3-24).

Non-ductile concrete frame structures, such as the warehouse buildings at Piers 7 and 8 of the Shinko Piers, suffered severe damage and collapse (Figures 3-25 and 3-26). The long, two-story warehouses along Piers 8A and 8B had a soft first story, consisting of a lower story (for parking) whose lateral force resisting system consisted of flexible exterior frames and some interior walls, and an upper story (for storage) whose lateral force resisting system consisted of stiff shear walls. Because of this, virtually all of the columns along the lower level of the buildings either collapsed or were severely damaged, and cranes mounted on the buildings overturned. A strong motion accelerometer in the upper level of one of these buildings recorded very strong motions (with a peak acceleration of about 700 cm/sec$^2$) whose duration of strong shaking was about 8 sec (Fig. 3-27). The durations of several of the peaks in the acceleration history were about 1 sec., indicating significant velocities and displacements of the structure.

### 3.6.2  Shear Wall Buildings and Newer Construction

Concrete shear wall warehouse buildings, such as those at Pier 6 of the Shinko Piers and at the Naka and Minami Wharves on Port Island, generally exhibited much better performance than did the above-indicated corrugated metal, wood, and non-ductile concrete frame buildings (although the interiors of these shear wall buildings were not accessible for observation) (Figure 3-28). In addition, modern buildings probably designed and detailed using more recent seismic design standards, such as the office buildings in the interior of Port Island and the passenger terminals at Port Island and Rokko Island, appeared to exhibit good seismic performance, even though there was often noticeable settlement of the soils along the exterior of the buildings (Fig. 3-22a). It is noted that the interiors of these passenger terminals were accessible for observation, whereas the office buildings were not; no damage to the interiors of the passenger terminals was observed at the time of our reconnaissance. The soil settlements alongside pile-supported buildings that otherwise performed well during the earthquake were larger at buildings located in the waterfront areas (i.e.; areas of lateral, as well as, vertical movement of the soils).

### 3.6.3 Parking Structure at Pier 4 of Shinko Piers

An interesting example of liquefaction-induced building damage at the port was a two-story parking structure at Pier 4 of the Shinko Piers. This structure is located just to the north of the south end of Pier 4 (see Fig. 3-29) where liquefaction of the fills caused lateral movement of the quay walls, substantial ground settlement, and lateral movement of the north support of the Kobe Ohashi Bridge (as discussed further below). The parking structure was founded on piles and pile caps interconnected by grade beams. Liquefaction of the fills beneath the structure led to extensive settlement (exceeding 2 m at some locations) of the pavement between the grade-beam/pile-cap supports (Figs. 3-30a through c). In addition, approach ramps leading from the roadway just north of the bridge into the parking structure were severely damaged (Figs. 3-29 and 3-30d and e). No significant damage to the aboveground structural elements of the parking structure was observed, suggesting that its pile foundation probably performed effectively (although the piles themselves were not accessible for observation during our reconnaissance).

## 3.7 Seismic Performance of Supporting Lifelines and Tanks

### 3.7.1 Highway Bridges

Highway bridges that service the Port of Kobe area also experienced damage that inhibited post-earthquake repair and reconstruction efforts at the port. Examples of this damage are described below:

o  *Kobe Ohashi Bridge.* This steel arch bridge that links Port Island to Pier 4 of the Shinko Piers and the City of Kobe experienced damage in the form of about 0.6 meters of horizontal movement of the base of its north column support relative to its pile cap support (Fig. 3-31). Liquefaction and extensive movement of the underlying soils was the major contributor to this observed damage. In addition, the earthquake damaged the north approach to the Kobe Ohashi Bridge, which extends along Pier 4 and is a reinforced concrete elevated viaduct with two-column bents. This approach suffered extensive column damage, due to inadequate shear reinforcement in the presence of strong ground shaking (Fig. 3-32). Both the Kobe Ohashi Bridge and its north approach were open to limited automobile traffic at the time of our reconnaissance.

o  *Maya Ohashi Bridge and the Dai-ni Maya Ohashi Bridge.* These two parallel adjacent bridges connect the western portion of the Maya Piers at Pier 1 to the eastern part of the Shinko Pier near Pier 8. Both bridges were damaged and closed to traffic during our reconnaissance. The Dai-ni Maya Ohashi Bridge suffered damage to its foundations, its columns (fracture of transverse reinforcement of columns and buckling of longitudinal column reinforcement), and its beam-column joints. Extensive soil movement at the base of the bridge columns was undoubtedly a significant contributor to this damage, in addition to ground shaking (Fig. 3-33). The Maya Ohashi Bridge is a cable-stayed structure which shifted off of its seats at the top of the piers (Fig. 3-34). Despite the movement of the main bridge deck, the support cables appeared to have retained most, if not all, of their tension.

o  *Nadahama Ohashi Bridge.* This bridge connects the northeast corner of the Maya Piers to the mainland. It appeared to be undamaged despite considerable liquefaction-induced soil deformation around the foundation.

o  *Hanshin Expressway Bridge (Osaka Bay Route).* This new steel arch bridge, which connects Rokko Island to the mainland, suffered a bearing failure due to excessive substructure movements (Fig. 3-35). This led to a racking of the arch and a buckling of the top cross framing. In addition to large inertia forces due to strong shaking, the bridge substructure was subjected to liquefaction-induced ground displacements (Fig. 3-35d). The results of detailed surveys of the bridge piers will indicate if the foundations have undergone permanent deformation, and will also yield important information on the interaction of this large pile-supported structure with its surrounding liquefied soils.

### 3.7.2  Rail System Damage

In addition to highway bridge damage along the Kobe Port, railroad bridges near the port also suffered damage (see Chapter 2). Information on the impact of this damage on the post-earthquake recovery of the Kobe Port is expected to become available in the months to come.

### 3.7.3  Utility Lifelines

As a result of the massive soil movement at the port, underground water, wastewater, and natural gas pipelines in the area were severely damaged. This damage had not yet been repaired at the time of our visit; as a result, service of these utility lifelines had not yet been restored, and there was no indication that these utility services would be restored in the immediate future. Power and telephone communication were operating at the port during our reconnaissance. Damage to these utilities was reportedly restored throughout the Kobe area (including the port) within a few days after the earthquake. Examples of damaged water lines at the Port of Kobe are shown in Figure 3-36.

### 3.7.4  Tanks

During our Kobe Port reconnaissance, we were able to carry out limited observations of tanks farms at the Nagata Harbor (western-most portion of the port) and near the entrance to the Fourth Reclamation Area (eastern-most portion of the port). The tanks at Nagata Harbor were observed to have tilted substantially, and minor wall buckling was observed at tanks in the Fourth Reclamation

Area (Fig. 3-37). More widespread damage at tank farms and also at oil and petrochemical terminals at the Kobe Port has been reported by others (e.g., Eskijian, 1995).

## TABLE 3-1
## HISTORY OF DEVELOPMENT AT THE PORT OF KOBE

| Pier | Dates of Reclamation | Dates of Construction | Total Length of Waterfront (m) | Total Area (Hectares) | Notes |
|---|---|---|---|---|---|
| Hyogo | Prior to 1900 | Piers 1-2, 1919-1939 Pier 3, 1958-1965 | 5,100 | 51 | Redevelopment of Piers 2 and 3, 1989-1993 |
| Takahama Wharf/Naka | Prior to 1900 | 1900-1920 | 3,600 | 34 | Redevelopment of piers, 1992-present |
| Shinko | Prior to 1900 | No. 1-4, 1896-1922 No. 4-6, 1919-1939 No. 7, 1951-1956 No. 8, 1954-1967 | 13,700 | 117 | -- |
| Maya | 1959 | 1959-1967 | 6,400 | 87 | Redevelopment of piers, 1987-1991 |
| Port Island Stage One | 1966-1981 | -- | -- | 436 | -- |
| Port Island Stage Two | 1987-1996 (est.) | -- | -- | 390 | -- |
| Port Island Total | -- | -- | 19,300 | 826 | -- |
| Rokko Island | 1972-1992 | -- | 12,200 | 580 | -- |
| Fourth Reclamation Area | 1934-1940 (mainland) 1965-1969 (island) | 1965-1970 | 3,900 | 51 | -- |

## TABLE 3-2
## GROUND CONDITION AND IMPORTANCE FACTORS USED TO COMPUTE SEISMIC DESIGN COEFFICIENT FOR PORT FACILITIES IN JAPAN
## (KANAI, 1983)

| Thickness of Quaternary Deposit | Gravel | Sand or Clay | Soft Ground |
|---|---|---|---|
| <5m | 1st kind | 1st kind | 2nd kind |
| 5-25m | 1st kind | 2nd kind | 3rd kind |
| >25m | 2nd kind | 3rd kind | 3rd kind |

a) Soil Classifications

| Classification | 1st kind | 2nd kind | 3rd kind |
|---|---|---|---|
| Factor | 0.8 | 1.0 | 1.2 |

b) Ground Condition Factor, G

| Structure | Characteristics of Structure | Factor |
|---|---|---|
| Special Class | The structure has significant characteristics described by items (1)-(3) in A class. | 1.5 |
| A Class | (1) If the structure is damaged by an earthquake, a large number of human life and property will possibly be lost. (2) The structure will perform an important role on the reconstruction work of the region after an earthquake. (3) The structure handles a hazardous or a dangerous object, and it is feared that the damage on the structure will cause a great loss of human life or property. (4) If the structure is damaged, economical and social activity of the region will be severely suffered. (5) If the structure is damaged, it is supposed that the repair work of it is considerably difficult. | 1.2 |
| B Class | The structure is other than Special, A and C classes. | 1.0 |
| C Class | The structure is small and easy for repairment, excepting that in Special and A classes. | 0.5 |

c) Importance Factor, I

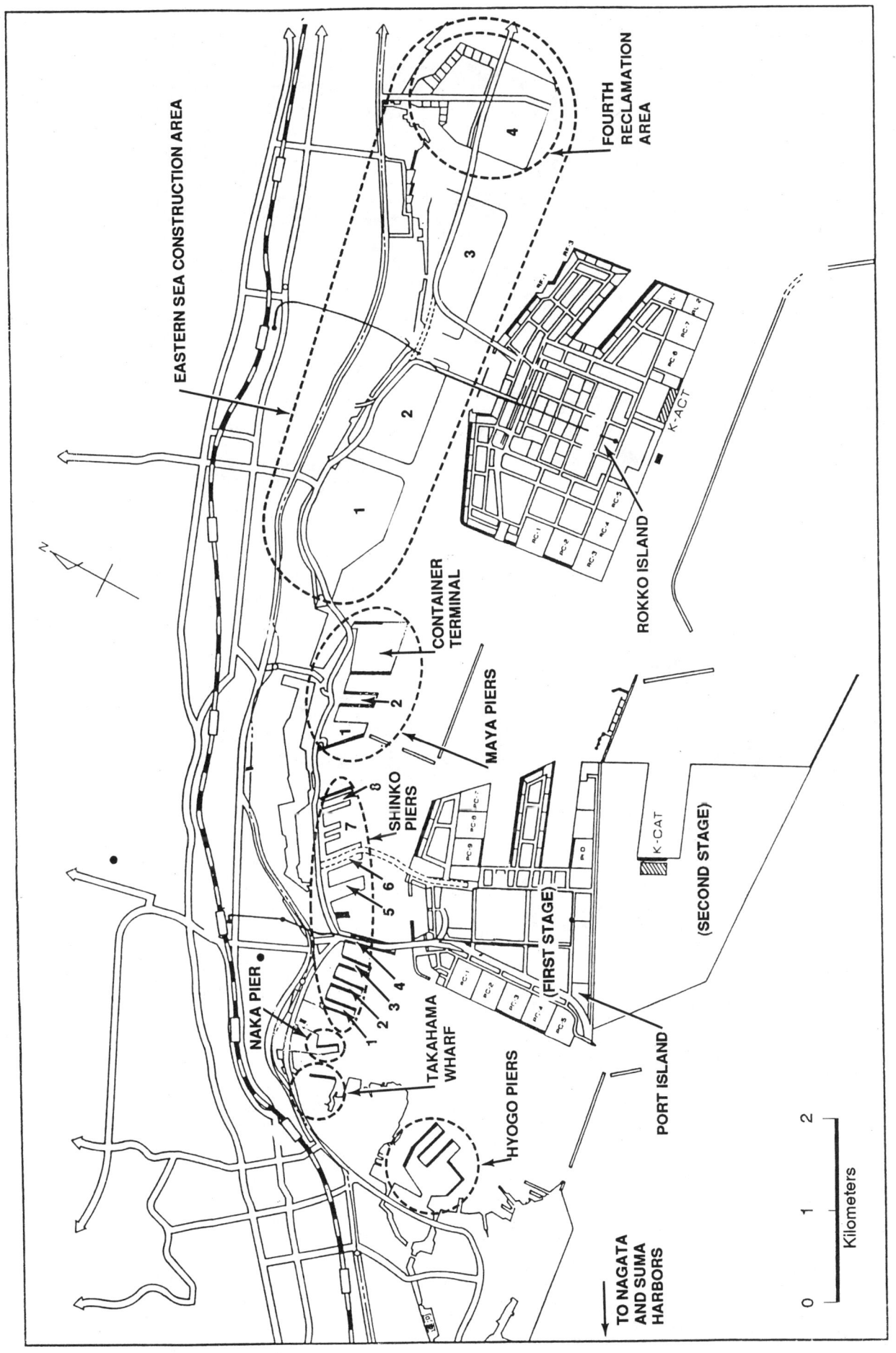

FIGURE 3-1
KOBE PORT

**a) Port Island and Kobe City to North and West**

**b) Rokko Island and Kobe City to North**

**FIGURE 3-2
AERIAL VIEWS OF PORT ISLAND AND ROKKO ISLAND
(Port of Kobe, 1994)**

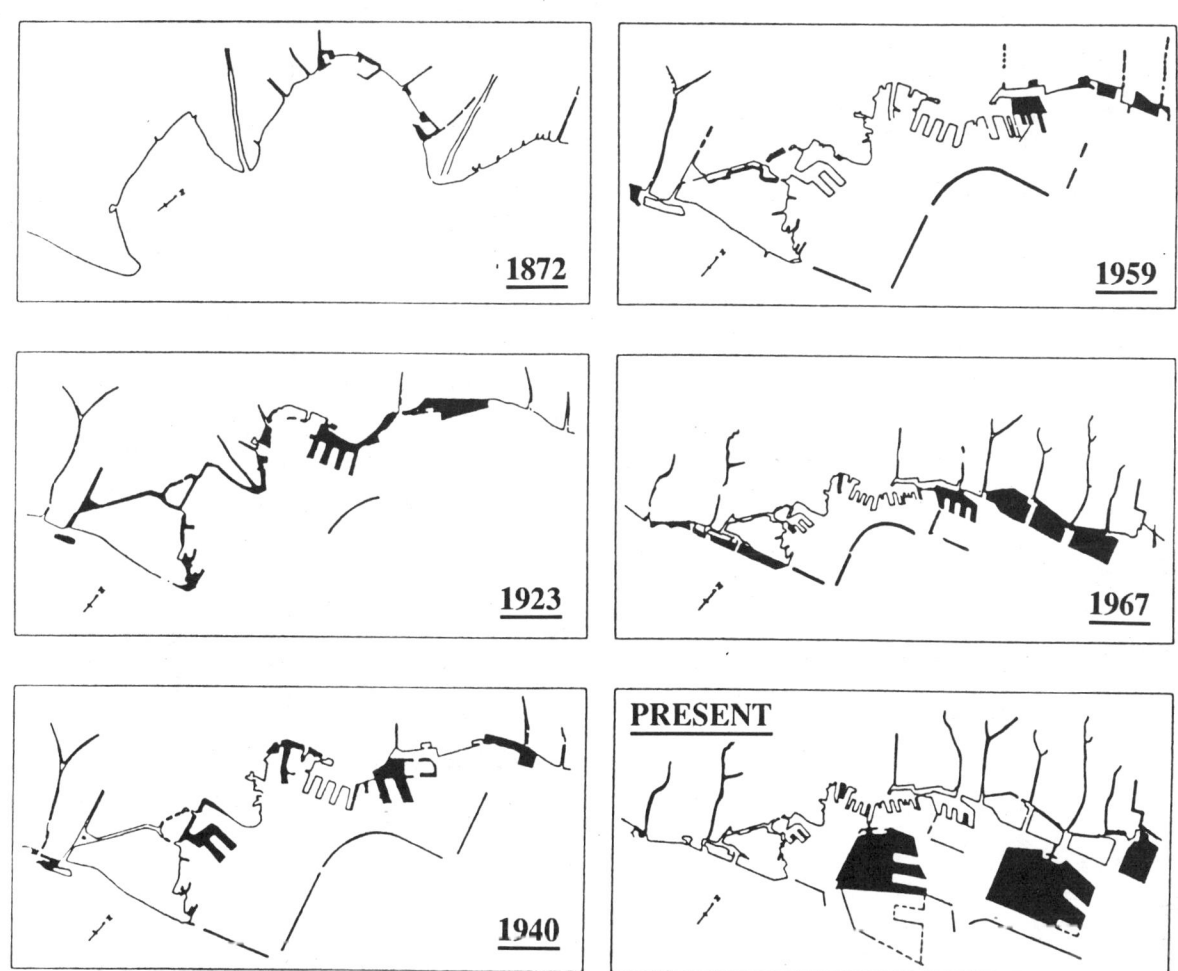

**FIGURE 3-3
DEVELOPMENT OF KOBE WATERFRONT (1872 TO PRESENT)**
(Kashima Construction Company, 1995)

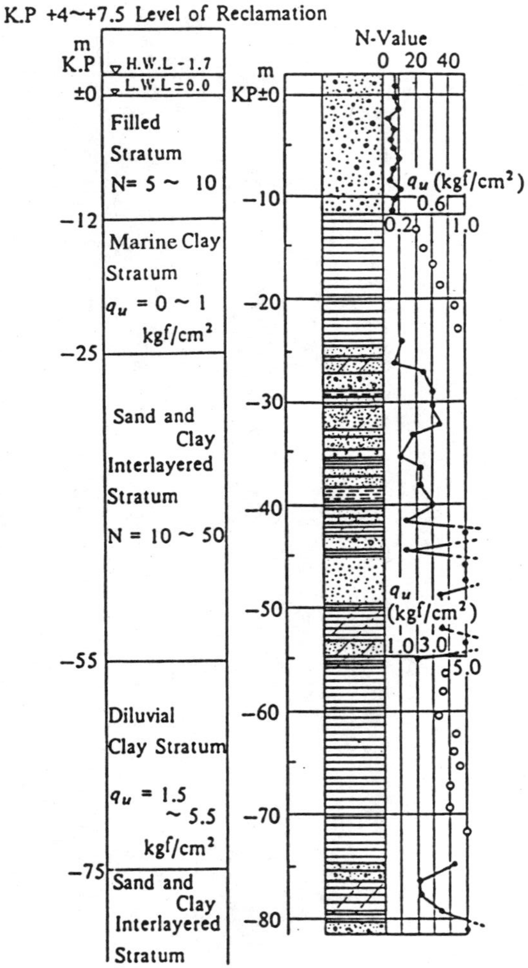

a) General Soil Profile at Port Island (First Stage)
(Nakakita and Watanabe, 1981)

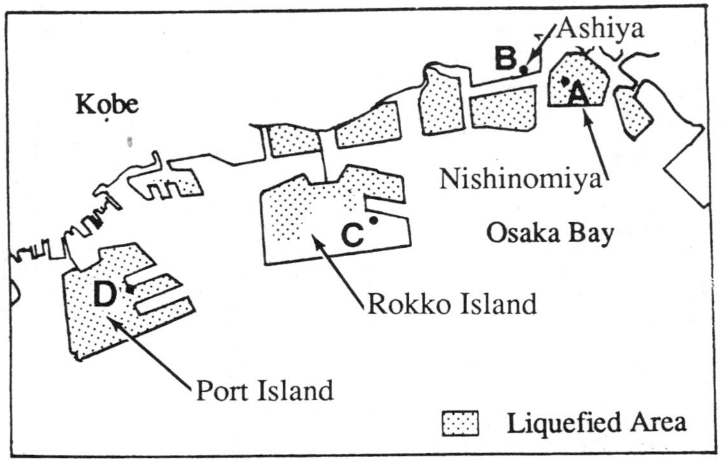

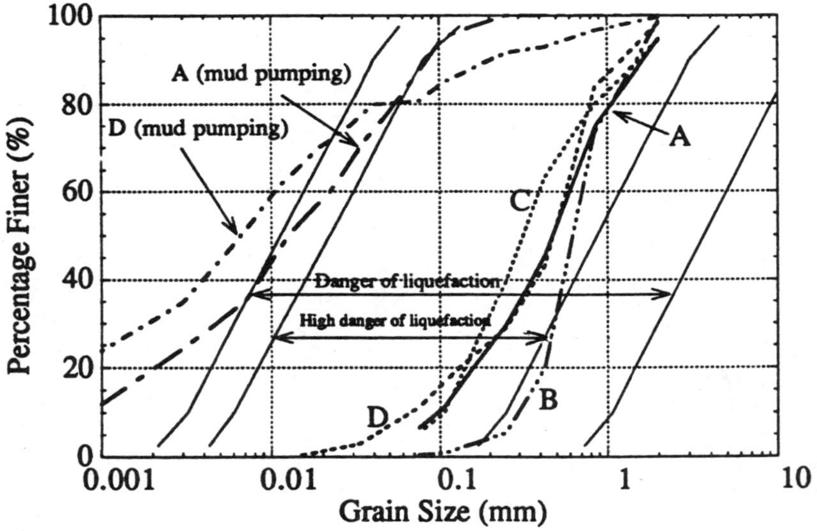

b) Grain Size Distribution of Sand Obtained from Liquefaction Features (Kamon et al, 1995)

**FIGURE 3-4
SOIL CHARACTERISTICS AT KOBE PORT**

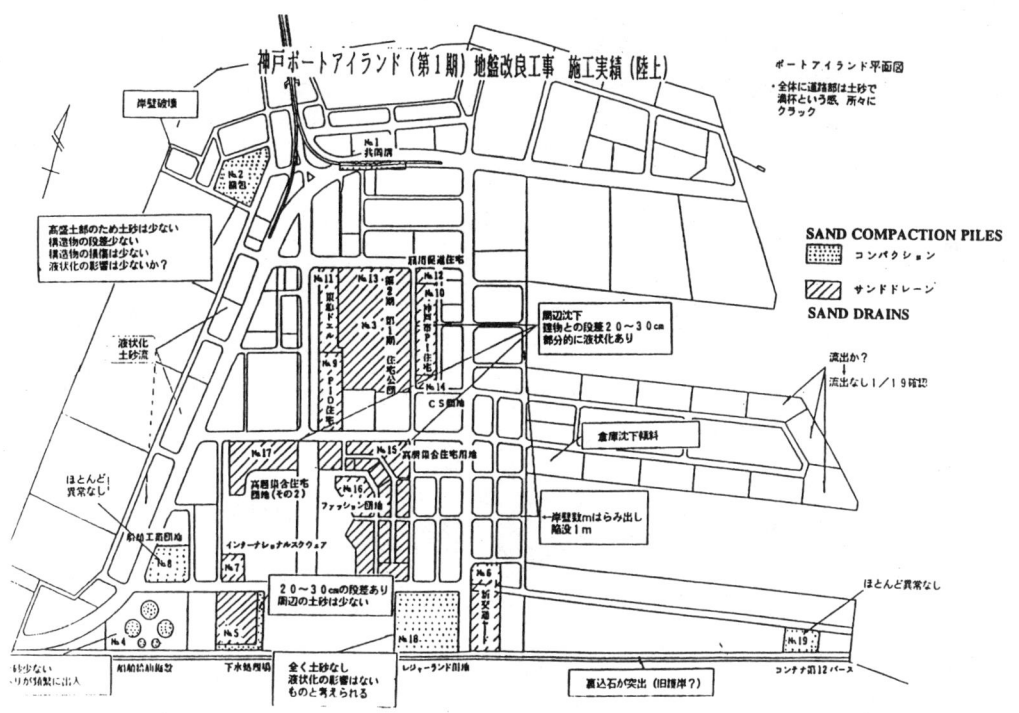

a) First Stage

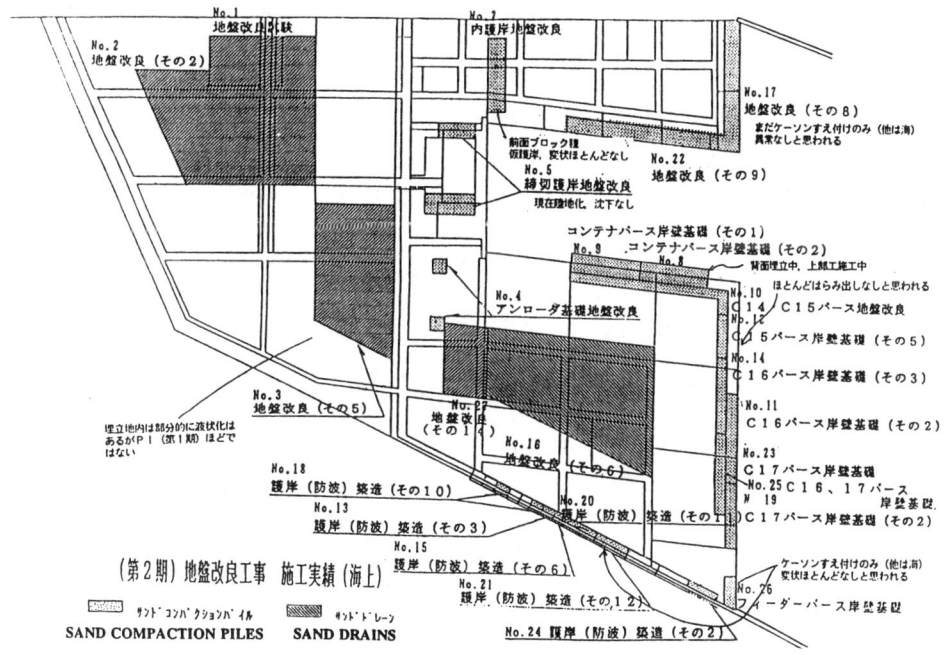

b) Second Stage

**FIGURE 3-5
SOIL IMPROVEMENT ON PORT ISLAND
(Fudo Construction Company, 1995)**

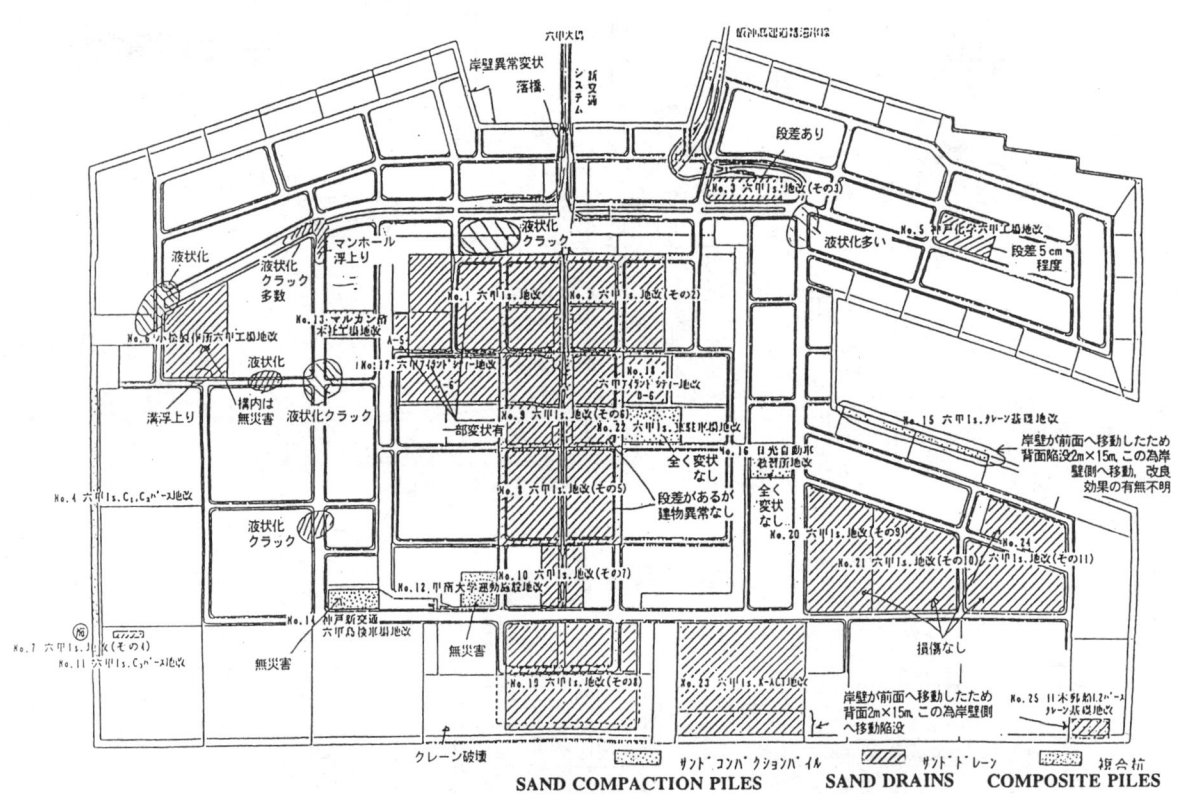

**FIGURE 3-6
SOIL IMPROVEMENT ON ROKKO ISLAND
(Fudo Construction Company, 1995)**

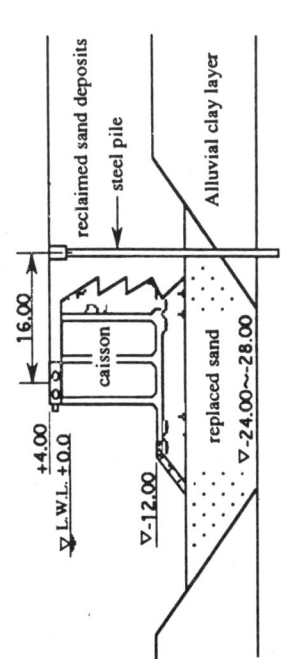

c) Replacement Caisson on Port Island
(Similar to Originals)

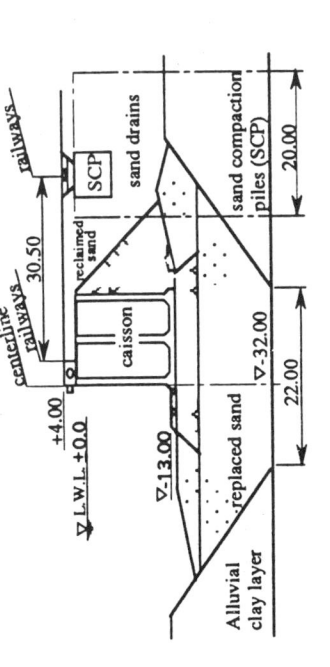

a) Quay of C1 and C2 Berth in Rokko Island*

b) Container Berth in Port Island*

\* Cross Sections from Kamon et al (1995)

**FIGURE 3-7
CAISSON QUAY WALL CROSS SECTIONS AT
PORT ISLAND AND ROKKO ISLAND**

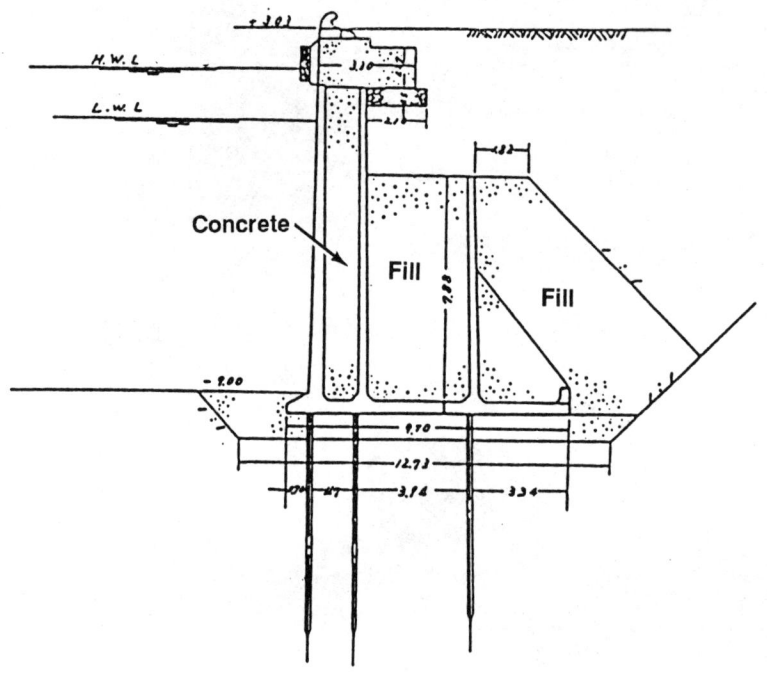

a) Deep Water Section, Naka Pier

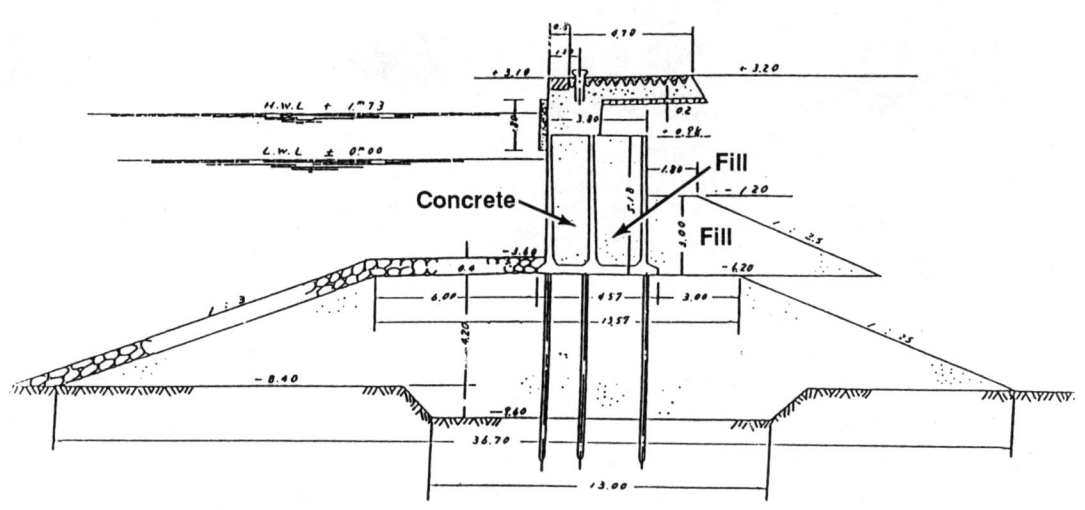

b) Shallow Water Section, Sixth Pier of Shinko Piers

**FIGURE 3-8
PILE-SUPPORTED CONCRETE CAISSON
QUAY WALLS AT KOBE PORT
(Iwasaki, 1995a)**

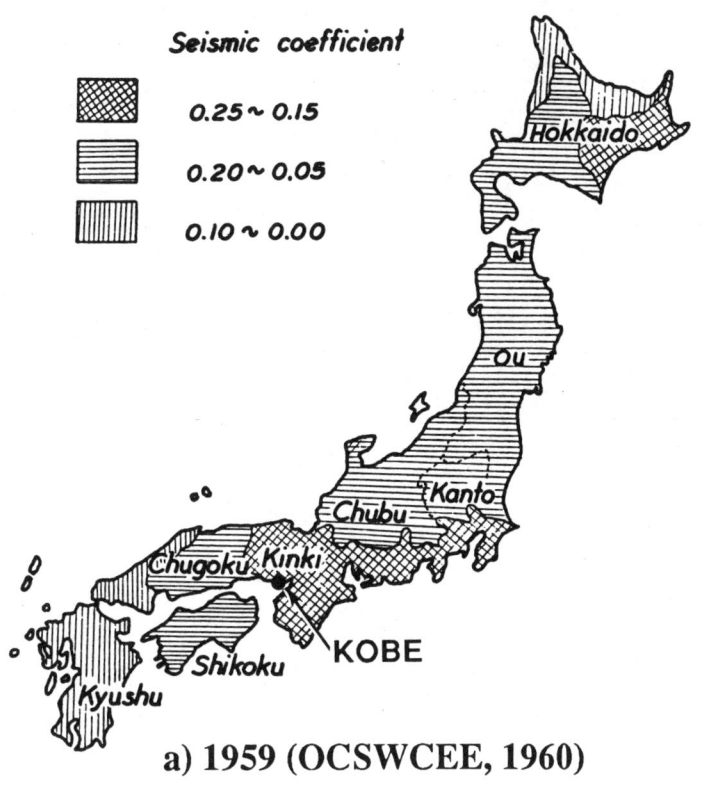

a) 1959 (OCSWCEE, 1960)

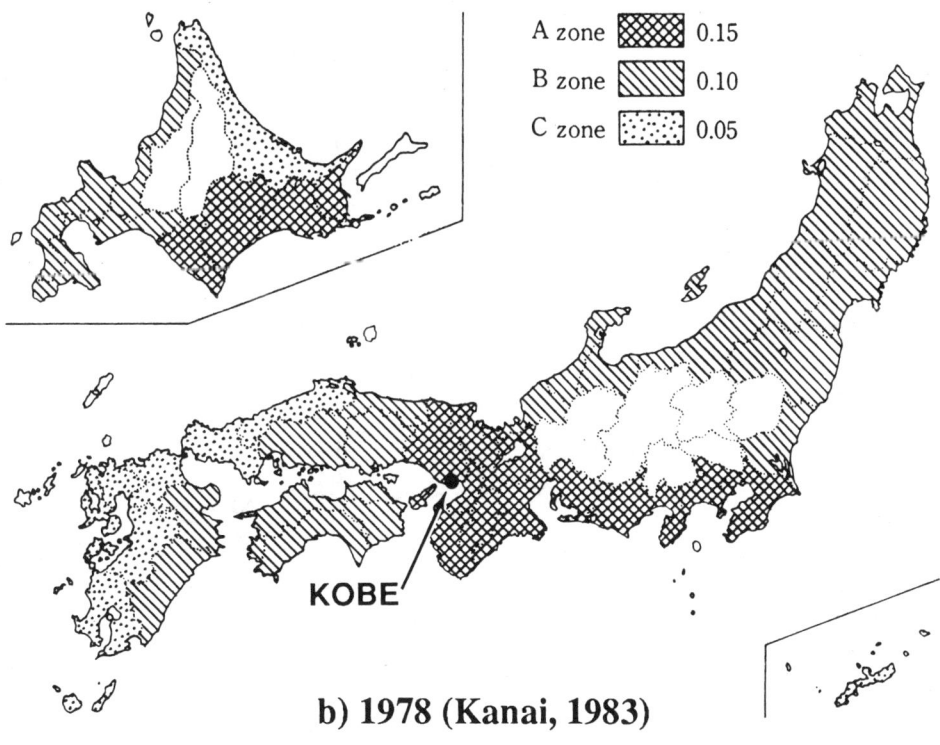

b) 1978 (Kanai, 1983)

**FIGURE 3-9
HORIZONTAL SEISMIC COEFFICIENTS
USED FOR DESIGN OF PORT FACILITIES IN JAPAN**

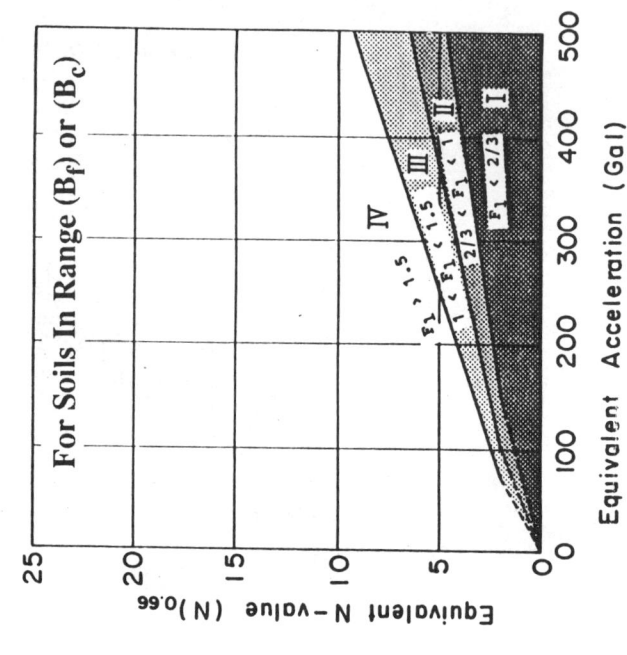

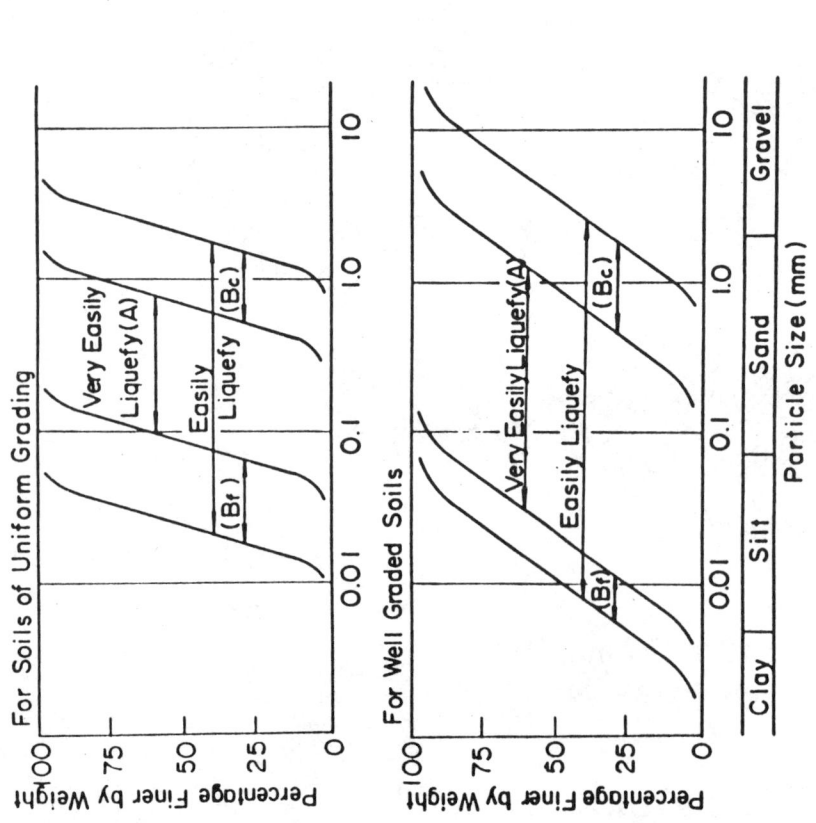

**FIGURE 3-10
ASSESSMENT OF LIQUEFACTION
POTENTIAL IN 1989 SEISMIC DESIGN
PROCEDURE FOR PORTS IN JAPAN
(Iai et al, 1989; Tsuchida 1990)**

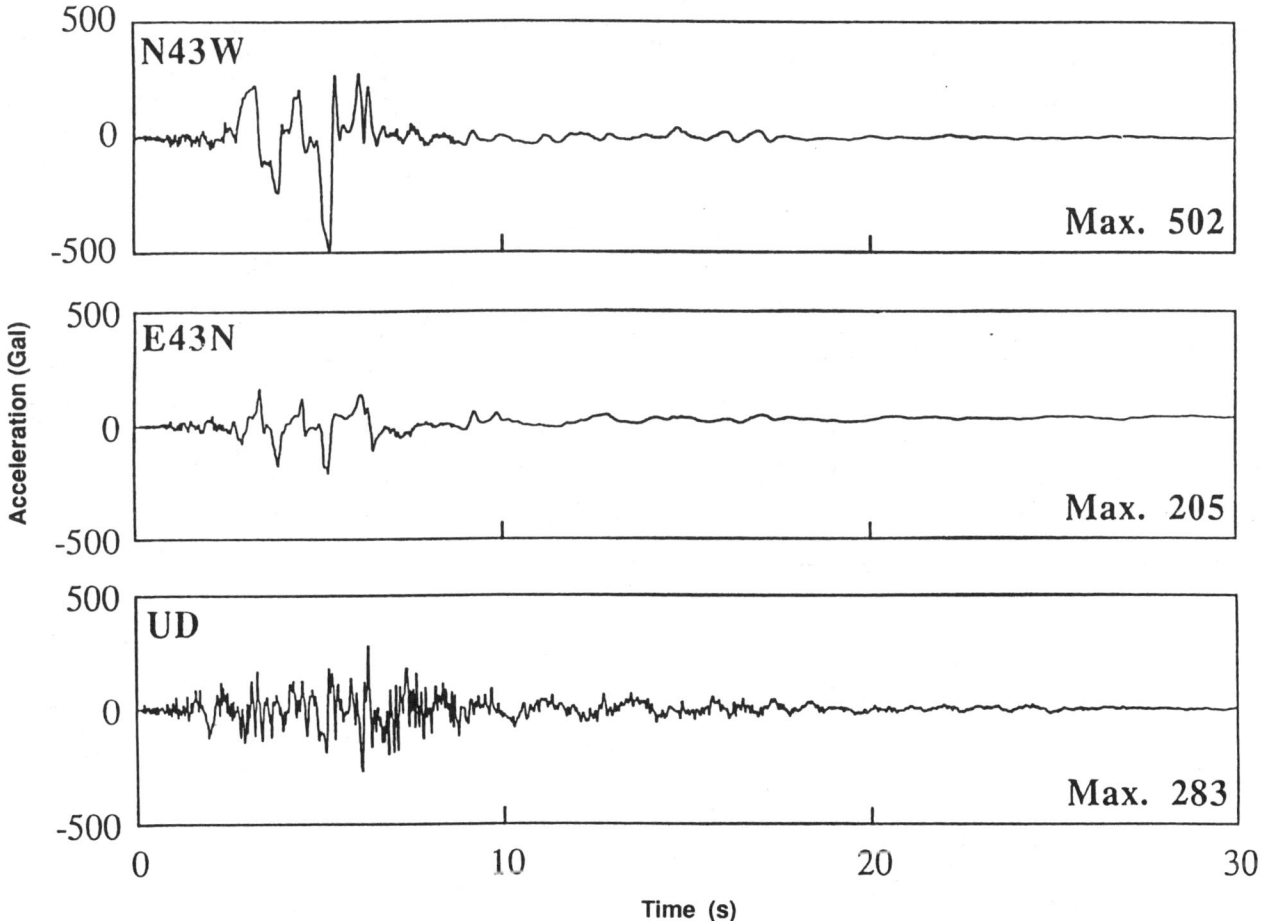

**FIGURE 3-11
GROUND MOTIONS RECORDED AT KOBE PORT
CONSTRUCTION OFFICE (ON MAINLAND)
(DPRI, 1995b)**

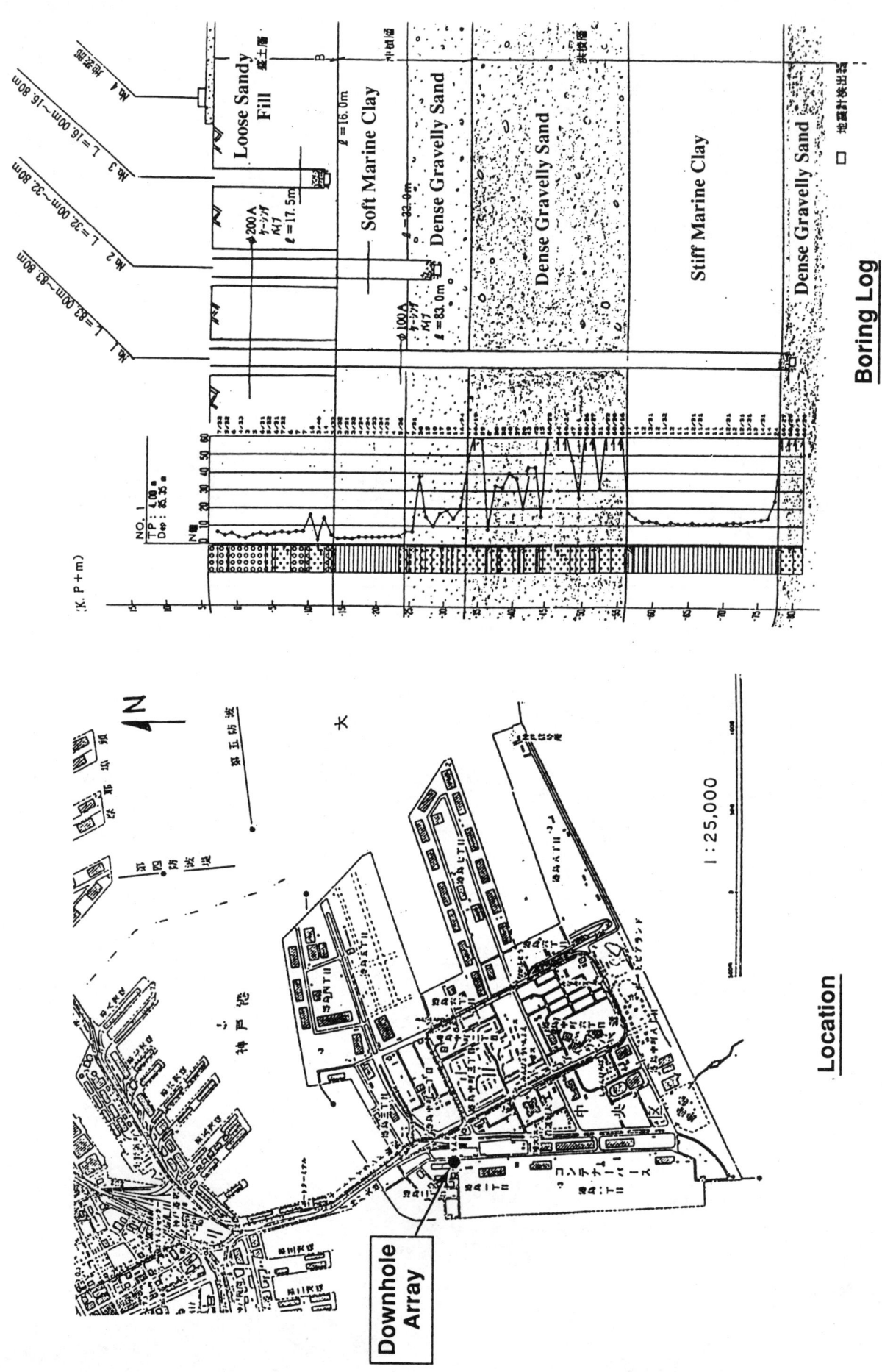

**FIGURE 3-12 DOWNHOLE ARRAY OF ACCELEROMETERS AT PORT ISLAND (GRI, 1995b)**

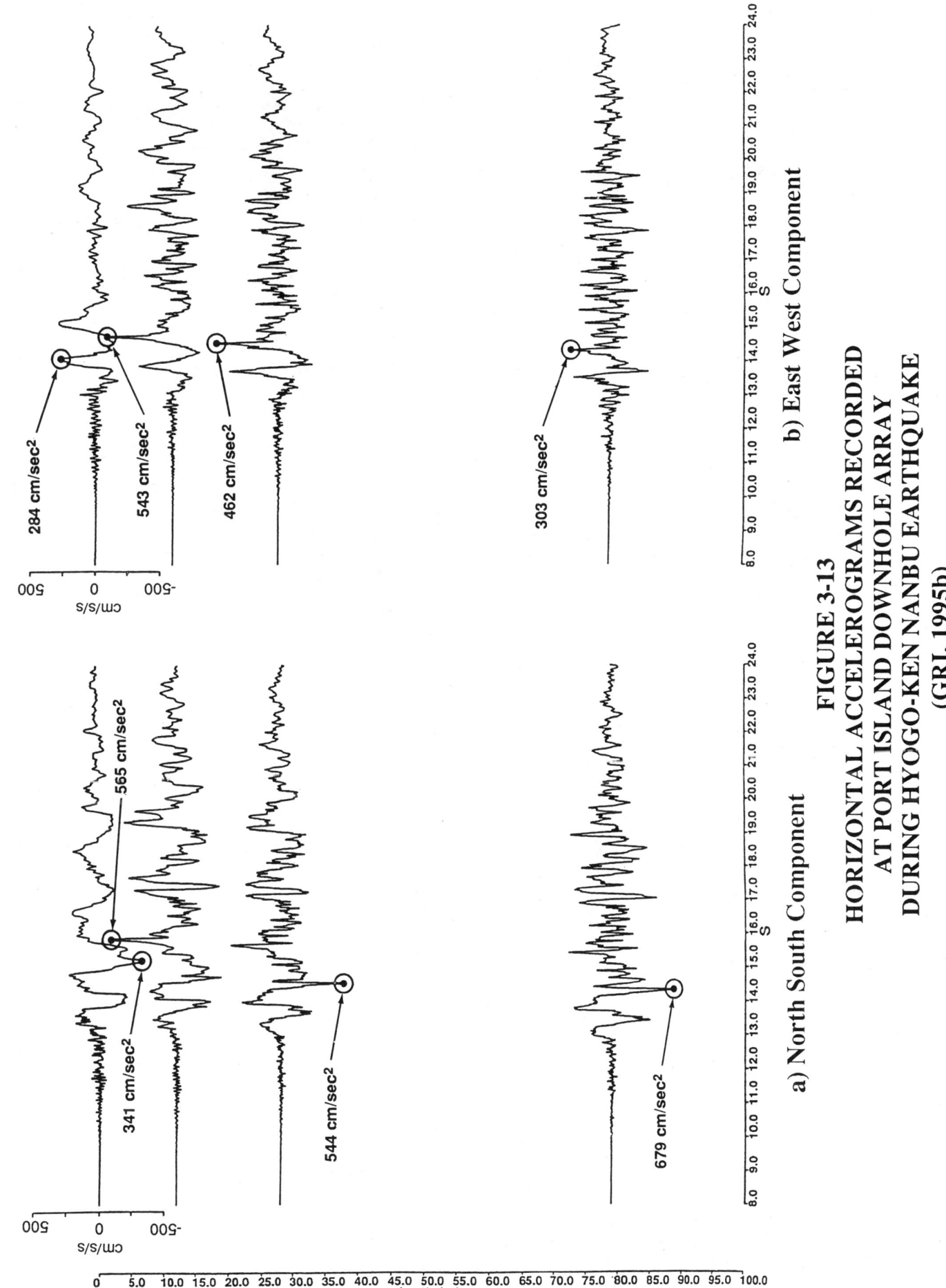

**FIGURE 3-13
HORIZONTAL ACCELEROGRAMS RECORDED
AT PORT ISLAND DOWNHOLE ARRAY
DURING HYOGO-KEN NANBU EARTHQUAKE
(GRI, 1995b)**

a) West Side of Port Island

b) Maya Container Terminal

**FIGURE 3-14
LIQUEFACTION OF FILLS AT KOBE PORT
(Aerial Photos of Port Island from
Kyodo News Agency, 1995 and
Asahi Newspaper Company, 1995b)**

a) Maya Container Terminal

b) Pier 1 at Port Island

c) Container Berths at Port Island

d) Rokko Island Ferry Terminal (JTL, 1995)

**FIGURE 3-15
QUAY WALL MOVEMENT AND FILL SETTLEMENT
KOBE PORT**

e) Pier 4 at Shinko Piers

f) North Side of Maya Piers
(Yomiuri Shimbun, 1995)

g) Hanky Ferry Terminal

h) Northeast Side
of Rokko Island

**FIGURE 3-15 (continued)**

i) Northeast Side of Rokko Island

j) On Mainland, between Naka Pier and Takahama Wharf

k) On Mainland just North of Pier 4, Shinko Piers

l) Hyogo Pier (Asahi Newspaper Co., 1995b)

FIGURE 3-15 (continued)

a) Buckled Legs, Rokko Island

b) Damaged Locking Mechanism, Port Island
(Photo by Dr. H.Tsuchida)

c) Collapsed Legs, Rokko Island

d) Closeup of (c)

FIGURE 3-16
CRANE DAMAGE AT KOBE PORT

e) Damaged Crane Rail at Minami Wharf, Port Island

f) Movement of Crane off of Rail, Rokko Island

g) Crane Repair, Rokko Island

h) Collapsed Support at Pier 7, Shinko Piers

**FIGURE 3-16 (continued)**

a) Looking North along East Face of Pier 1, at Location of Soil Settlement Adjacent to Steel Cell Caissons

b) Close-up of Soil Movement from a) Above

FIGURE 3-17
SEISMIC PERFORMANCE OF STEEL PLATE CELLULAR BULKHEADS AT PIER 1 OF MAYA PIERS

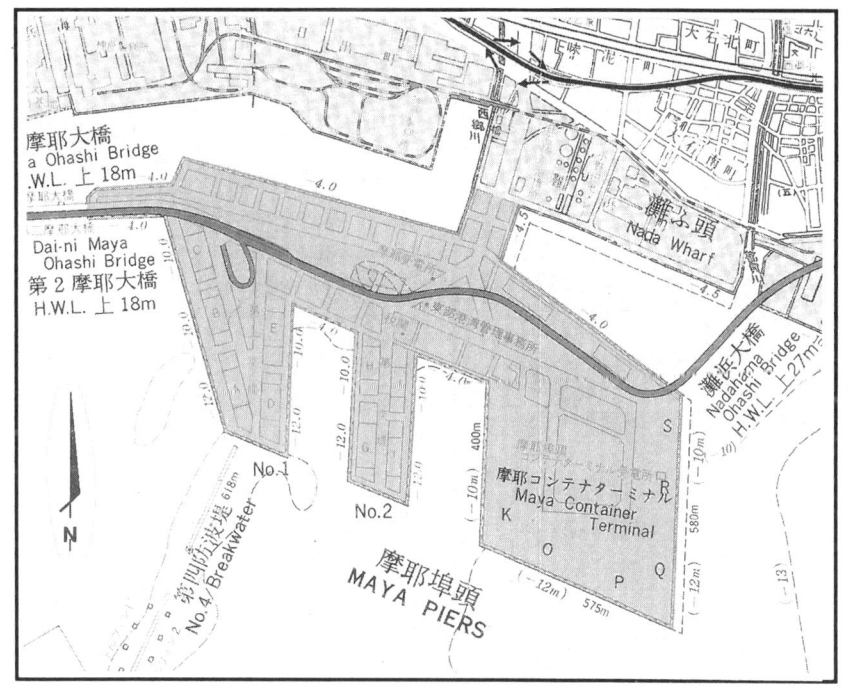

**FIGURE 3-18
MAYA PIERS AND CONTAINER TERMINAL
(Port of Kobe, 1994)**

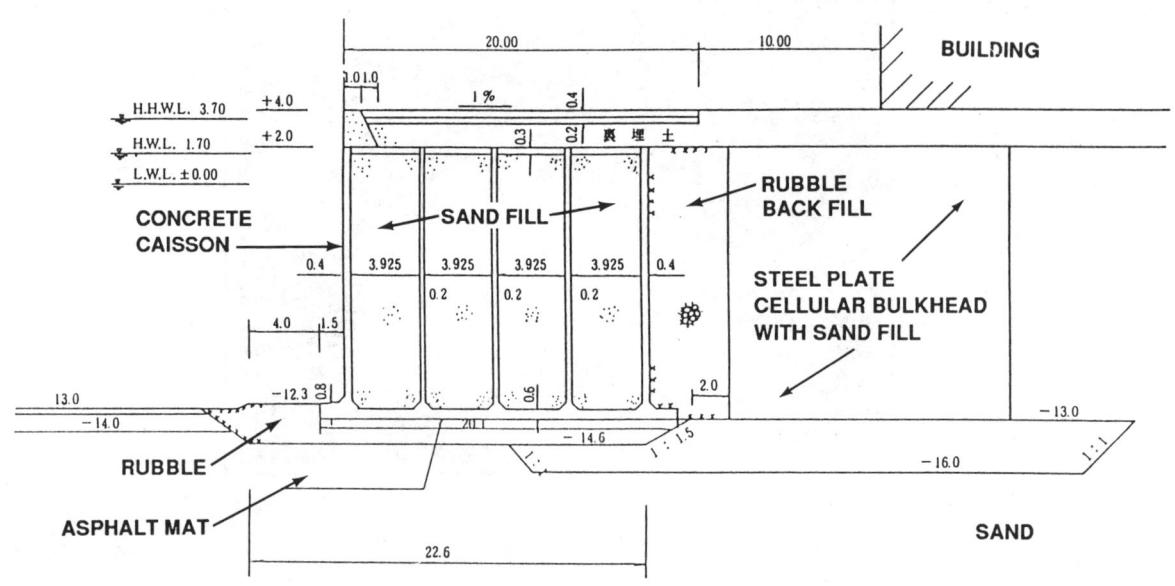

a) Berth A

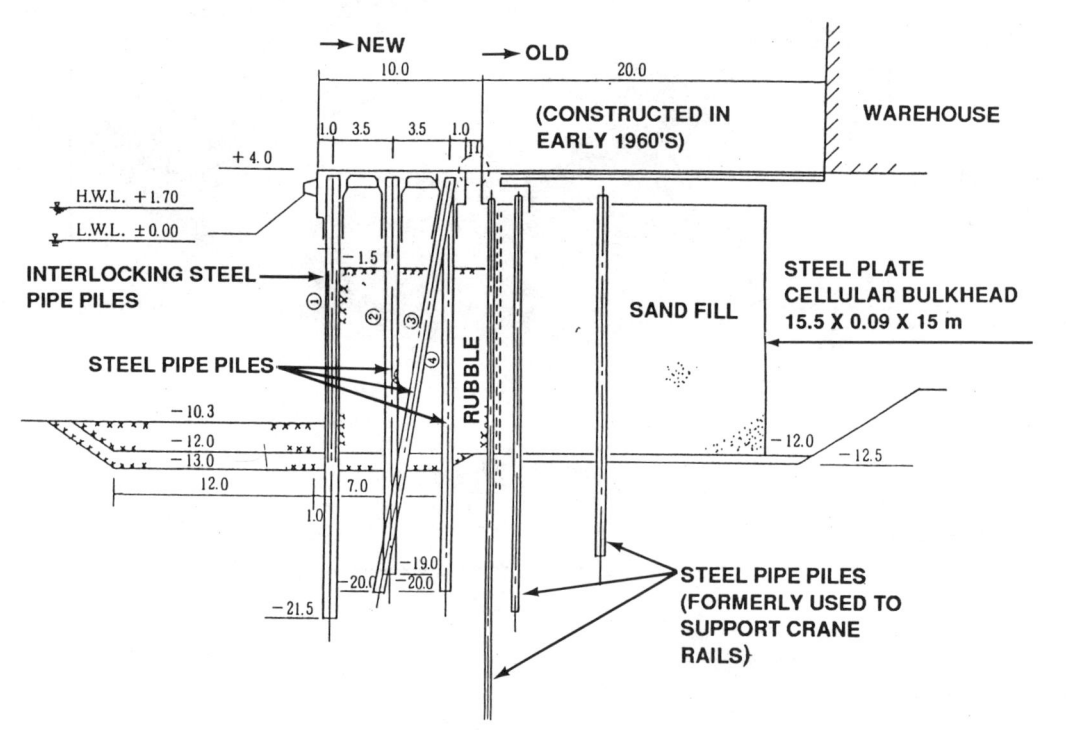

b) Berth C

**FIGURE 3-19
CROSS SECTION OF WATERFRONT RETAINING
STRUCTURES AT MAYA PIER 1 (dimensions in meters)**

a) Movement of Quay Wall at North Approach to the Maya Piers (Same Location as Fig. 3-15f)

b) Damage to Lightly Reinforced Concrete Cylinder Piles Beneath Building in Foreground of a).

c) Lateral Movement of Caissons at Berth O.

d) Area Between Piers 1 and 2

**FIGURE 3-20
SEISMIC PERFORMANCE OF WATERFRONT
STRUCTURES AT THE MAYA PIERS**

e) Berth E, Pier 1 (Looking Toward Area in d)

f) Southern End of Pier 1

g) Southwestern end of Pier 1

h) View Along West Edge of Pier 1 (Looking North from Berth A)

FIGURE 3-20 (continued)

3-44

i) Minor Quay Wall Damage between Berths B and C

j) Relative Seaward Movement between Berth B (background) and Berth C (foreground)

k) Waterfront at Berth C

l) Interlocking Steel Pipe Piles at Berth C

**FIGURE 3-20 (continued)**

a) Buckling of Steel Pipe Pile -
Fourth Reclamation Area
(Photo Courtesy of Mitsui Construction Company)

b) Fracture of Hollow Concrete
Cylinder Pile at Pile Cap -
Port Island, Pier 1

FIGURE 3-21
PILE DAMAGE AT KOBE PORT

a) View of Wharf Looking South

b) Wharf Surface after Earthquake

**FIGURE 3-22
GOOD SEISMIC PERFORMANCE OF A PILE-SUPPORTED
WHARF (TAKAHAMA WHARF)**

a) Commercial Building - Interior of Port Island

b) Commercial Building - Interior of Port Island

c) Port Building - Fourth Reclamation Area

d) Monorail - Port Island

**FIGURE 3-23
GROUND SETTLEMENT ADJACENT TO
PILE SUPPORTED STRUCTURES**

a) Connection Failure of Braced Corrugated Metal Warehouse at Naka Wharf, Port Island

b) Door Damage at Building in a)

c) Floor Settlement at Building in a)

d) Damage to Wooden Warehouse, Hyogo Pier

FIGURE 3-24
DAMAGE TO CORRUGATED METAL AND
WOODEN BUILDINGS AT KOBE PORT

**FIGURE 3-25
POOR PERFORMANCE OF NON-DUCTILE CONCRETE
FRAME WAREHOUSE STRUCTURE AT
PIERS 8A AND 8B, SHINKO PIERS**

**FIGURE 3-26
DAMAGE TO NON-DUCTILE CONCRETE FRAME WAREHOUSE
STRUCTURE AT PIER 7, SHINKO PIERS**

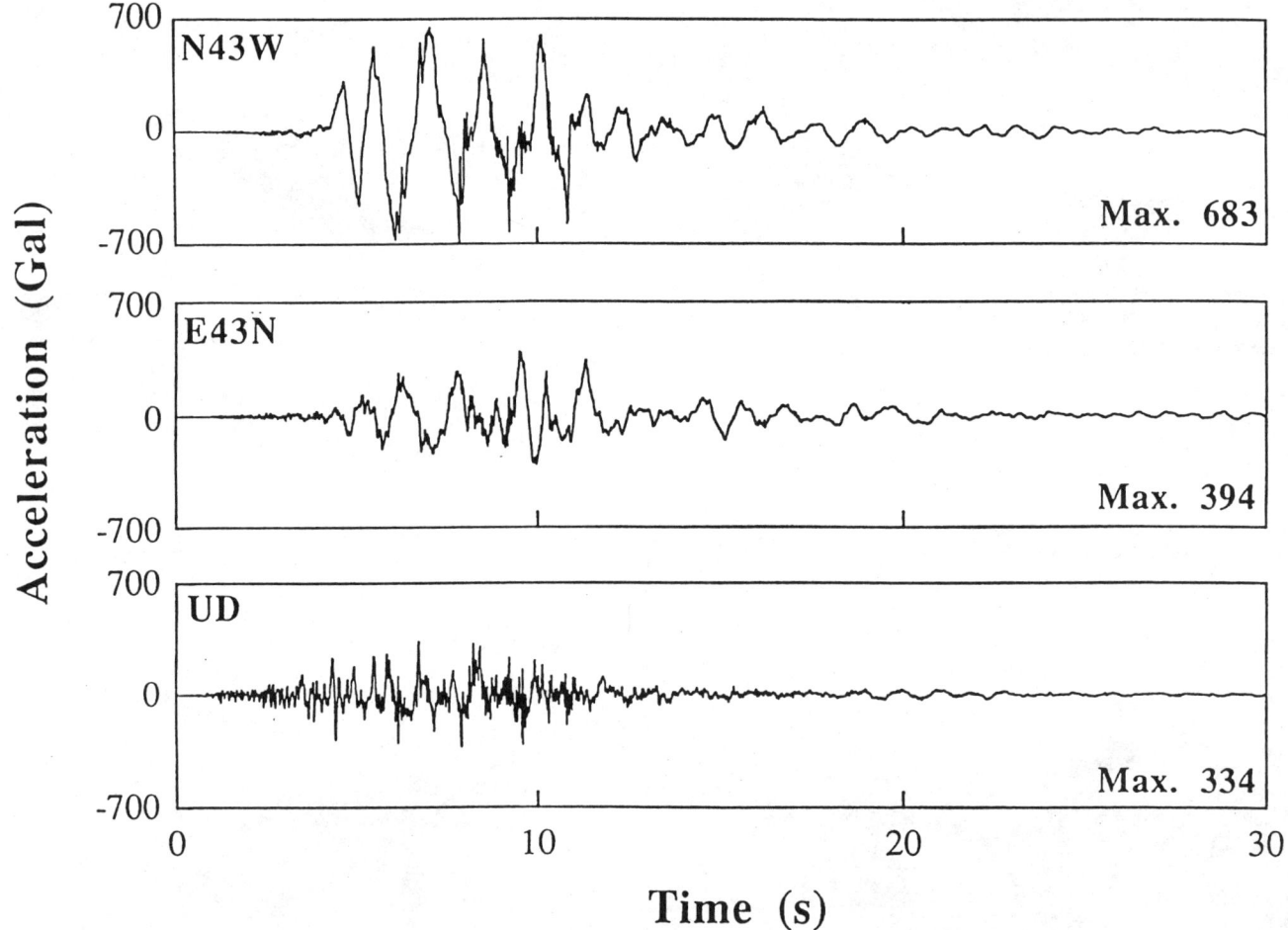

**FIGURE 3-27
STRONG MOTION ACCELEROGRAM
RECORDED AT PIER 8, SHINKO PIERS
(DPRI, 1995b)**

**FIGURE 3-28
GOOD SEISMIC PERFORMANCE OF
CONCRETE SHEAR WALL STRUCTURE AT
PIER 6, SHINKO PIERS**

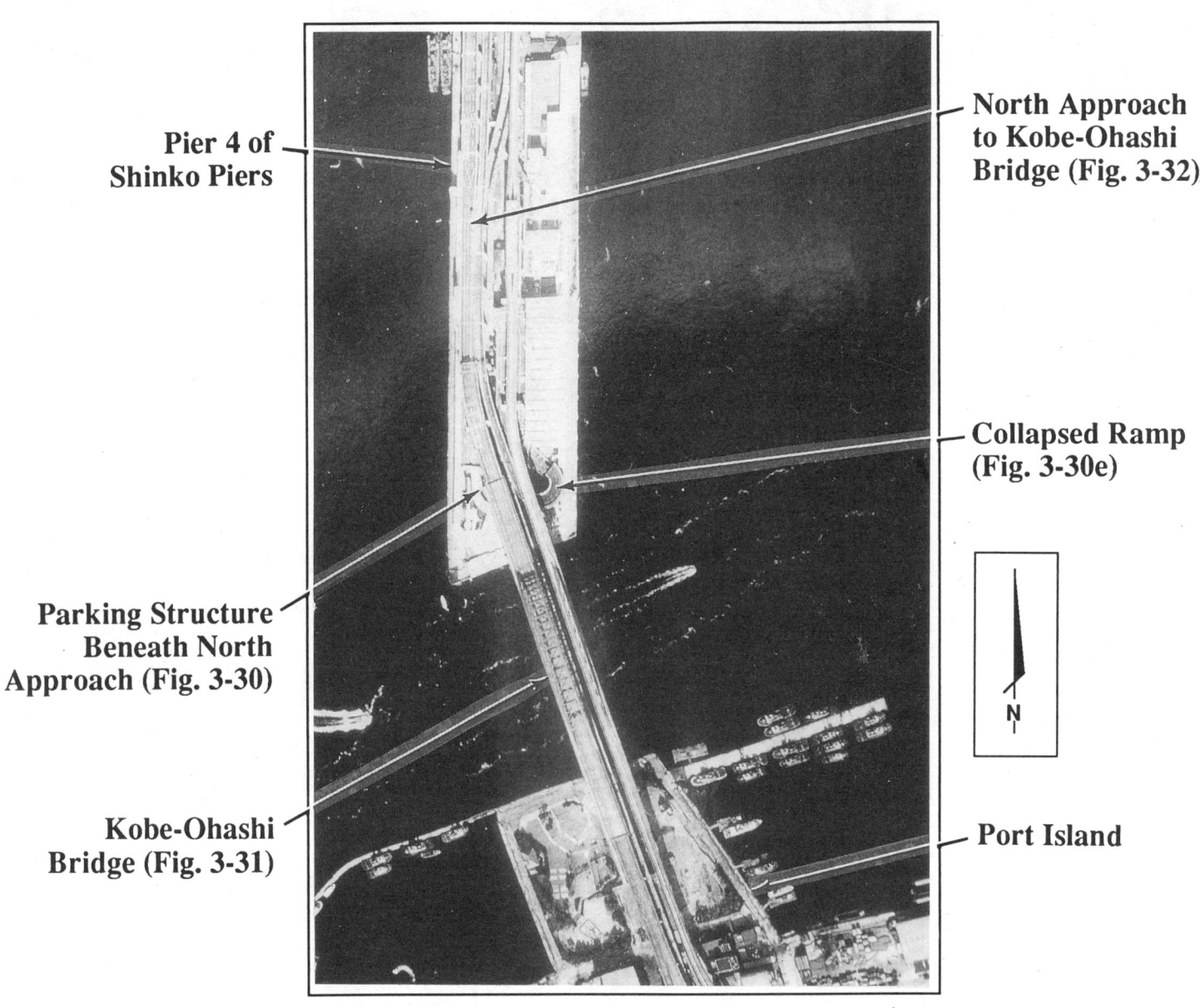

**FIGURE 3-29
AERIAL VIEW OF PIER 4 AND
KOBE-OHASHI BRIDGE AFTER EARTHQUAKE**
(Mainichi Newspaper Company, 1995)

a) View Looking South

b) Fill Settlement

c) Fill Settlement

d) Damaged Ramp

e) Collapsed Ramp

**FIGURE 3-30
DAMAGE TO PARKING STRUCTURE AND ADJACENT BRIDGE RAMPS
PIER 4, SHINKO PIERS**

a) Quay Wall Damage

b) Movement of North Support 2ft (0.7m) to North

c) Ground Movement

d) Ground Movement

**FIGURE 3-31
DAMAGE TO KOBE-OHASHI BRIDGE BETWEEN
MAINLAND AND PORT ISLAND**

a) Looking South Toward Port Island

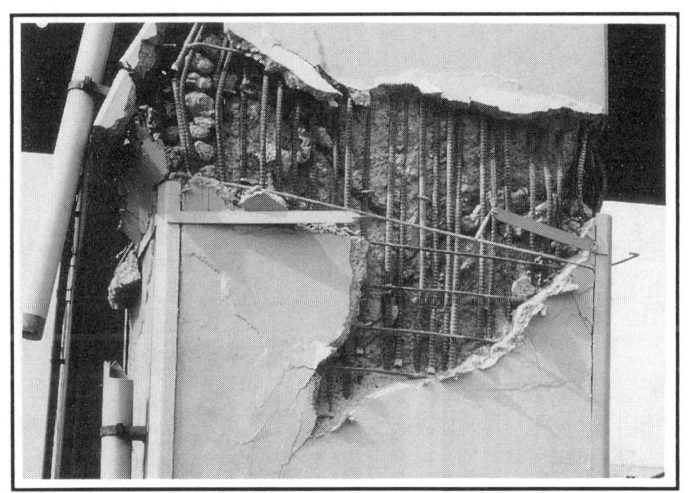

b) Column Damage - Inadequate Transverse Reinforcement

**FIGURE 3-32
DAMAGE TO NORTH APPROACH TO
KOBE-OHASHI BRIDGE**

a) Ground Movement

b) Repair to Damage at Beam-Column Joint

c) Column Damage

d) Close-up of Column Damage from c)

**FIGURE 3-33
DAMAGE TO DAI-NI MAYA OHASHI BRIDGE
(BETWEEN MAYA PIERS AND SHINKO PIERS)**

**FIGURE 3-34
DAMAGE TO MAYA-OHASHI BRIDGE -
LATERAL MOVEMENT OF DECK OFF OF BEARING SUPPORTS
(BETWEEN MAYA PIERS AND SHINKO PIERS)**

a) View Looking East (Hanshin Expressway Bridge in Background, Rokko Ohashi Bridge in Foreground)

b) Lateral Deformation of Superstructure and Bearing Failure

c) Closeup of Bearing Failure from b)

d) Ground Deformation at South End of Bridge (on Rokko Island)

**FIGURE 3-35
DAMAGE TO HANSHIN EXPRESSWAY, OSAKA BAY ROUTE
(STEEL ARCH BRIDGE CONNECTING
MAINLAND TO ROKKO ISLAND)**

a) Potable Water Line to Rokko Island

b) Maya Piers

c) Pier 4, Shinko Piers

**FIGURE 3-36
DAMAGED WATER LINES AT PORT OF KOBE**

a) Tank Tilting, Nagata Harbor

b) Buckling of Walls, Fourth Reclamation Area

**FIGURE 3-37
TANK DAMAGE AT KOBE PORT**

# CHAPTER 4
# OSAKA PORT

## 4.1 General Background

### 4.1.1 Port Description

The Port of Osaka is located at the northeastern edge of Osaka Bay, approximately 25 km west of downtown Kobe (Fig. 4-1). The port is one of Japan's largest, handling $9.23 \times 10^7$ metric tons of cargo through 185 berths (173 mooring facilities and 12 container berths) in 1993. In 1995, this total is expected to exceed $1.00 \times 10^8$.

The port complex includes: (a) an extensive network of mainland wharves and terminals built on reclaimed land along several rivers; (b) the 1002 ha South Port area which has 12 container cargo terminals, numerous ferry terminals and wharves, business zones, and residential and recreational areas; and (c) the North Port districts (North and South) which are now under construction, will add an additional 615 ha of similar facilities to the Port of Osaka (Fig. 4-2).

A total of 185 berths are available for foreign and domestic trade. These facilities include 12 container berths, liner terminals for freighters, numerous passenger ferry berths, intermodal terminals, and various wharves for the handling of food, lumber products and chemicals. In addition to the facilities devoted to marine commerce, the South Port and Tempozan District are centers for residential and commercial development as well as, recreational attractions. Lifeline infrastructure at the port includes an extensive system of elevated highway structures and numerous moderate-sized bridges (main spans of 100 m to 510 m), and an oil-fired power plant. Construction has also begun on an extension to the existing subway system. This development includes the Osaka Nannkou Tunnel which will connect the Tempozan District with the North Port reclamation areas.

### 4.1.2 Chronology of Port Development

The Port of Osaka has grown in much the same fashion as the Port of Kobe, and these similarities make it possible to compare (with due consideration for the intensity of ground motions) the seismic performance of the two facilities during the Hyogo-Ken Nanbu Earthquake. The similarities in the facilities and infrastructure at the Ports of Kobe and Osaka are evident in aerial photographs of the latter port (Fig. 4-3).

The port was established in 1868 and large scale reclamation and port construction was initiated in 1897. It is built entirely on fill, and the bayside reclamation initially involved end-dumping from land and, later, barge-dumping soil onto the soft marine clay found along the margins of Osaka Bay. The Port and Harbor Bureau of the City of Osaka currently manages port and harbor facilities which occupy approximately 5,600 ha and mainland waterfront districts comprising roughly 1,670 ha. The development plan of the inner portions of the reclamation tracts is also very similar to that carried out at the Port of Kobe, in that these areas include extensive commercial, residential, and recreational facilities, as well as "green belt" open-space areas (Fig. 4-2). The Port and Harbor Bureau (1990) estimates that, at the completion of current reclamation and development of "Technoport Osaka" in 2005, the North and South Port areas will have a resident population of 60,000 and a daytime population of 200,000. The magnitude of the projects is demonstrated by capital expenditure figures provided by the Port and Harbor Bureau (1990) for land reclamation and construction of infrastructure (public investment of ¥900 billion) and for the construction of commercial and residential facilities (private investment expected to reach ¥1.3 trillion).

The port has been subdivided into three primary sections; the mainland area, the South Port area, and the North Port area (Fig. 4-2). Almost all of the mainland portion of the port was reclaimed prior to World War II. The port facilities in this area were destroyed during the war, and the late 1950's ushered in a period of intensive modernization and development that has continued to the present day.

The South Port reclamation project was initiated in 1958 and in 1990 was 93% completed. The shallow water depths adjacent to the South Port facilities necessitate continuous dredging in shipping lanes. The handling of large volumes of dredge spoils, combined with the lack of local sources for granular fill soils, led to the utilization of the bay-floor sediments (predominantly fine-grained soils) as fill throughout the South Port region. This is in contrast to the sandy soils used in the Port of Kobe reclamation projects. The reclamation of the South Port area was performed in two stages. Stage One, carried out from 1963-1970, was directed toward improving the marine clays which are generally 15 to 25 m thick beneath in the South Port area. This soil improvement included the use of sand drains and pumping wells to increase the rate of consolidation of the marine clays as filling commenced. The second stage of reclamation focussed on improving the ground that was formed by filling with clayey soils. This was accomplished using cardboard wick drains and preload fills. The construction practices and settlements experienced in the South Port area have been well documented by Mikasa and Ohnishi (1981). During our post-earthquake reconnaissance, port engineers explained that the use of the clayey fill has resulted in widespread settlement problems beneath concrete caissons and crane rails.

The two newer portions of the Port of Osaka that are currently under construction comprise the North Port. The North Port area is subdivided into the North Port North District and the North Port South District (Fig. 4-2). When completed, the North Port will cover an area of 775 ha, 115 ha of which will be devoted to distribution and port terminals. The North Port reclamation was initiated with the placement of concrete caisson breakwaters and dikes in the North District in 1972. The fill used in the North Port reclamation is composed of bay-floor sediments, dredge spoil, and a significant proportion of solid municipal waste. Port representatives estimate that approximately 80% of the fill material includes excavated or dredged soils and debris from construction sites, and approximately 20% contains solid waste (e.g.; sludge, industrial and residential waste) from the city of Osaka. In 1990, the Port and Harbor Bureau estimated that by 1995 approximately $7.5 \times 10^7 \text{ m}^3$ of waste would be incorporated into the North and South District reclamation projects. The environmental issues associated with the inclusion of industrial and residential waste in the fill has resulted in the construction of several different waterfront retaining systems and the use of soil improvement techniques to minimize the potential for damage to the retaining/containment systems during earthquakes. The reclamation work was completed in the North District in 1985, and the work at the South District of the North Port will continue for several more years.

### 4.1.2 Soil Conditions

The City of Osaka has been constructed on reclaimed land at the margins of the delta formed by the Yodo River. The Port of Osaka is bounded to the north by the Yodo River and is bisected by four other rivers before reaching the Yamato River, which forms the southern boundary of the port. The soil units underlying the port complex are fairly uniform, and are similar to those at the Port of Kobe. The soil conditions in the South Port generally consist of 15 to 20 m of clayey fill underlain by 15 to 20 m of very soft to soft marine clay; 10 m of interbedded very loose clayey sand and soft sandy clay; and a dense to very dense gravelly sand to the depth of explorations at 40 m. A geotechnical boring log showing the typical profile before reclamation in the South Port is shown in Figure 4-4. The soil profiles in the North Port are very similar, although the marine clay deposit is generally thicker as the distance from the shore increases.

Foundation support for most of the larger structures in the port is provided by end-bearing piles which are embedded into the dense granular soils. The depth to bedrock (defined as material with a shear wave velocity greater than 800 m/sec) is approximately 700 m at the port (Iwasaki et. al., 1994).

The reclamation process in the South Port was initiated in 1963 by placing a 1.5 m thick sand mat directly over the clayey bay floor sediments, and then filling with predominantly cohesive dredge spoils. As outlined by Mikasa and Ohnishi (1981), the first phase of soil improvement focused on reducing the time required for the consolidation of the native marine clay under the loads imposed by the fill. A network of 40 cm diameter sand drains was installed in the marine clay. The drains extended to depths of 19 m below the original mudline and terminated in the sand mat, which was confined between the native clay and the clayey fill soils. A system of wells and pumps were then used to de-water the sand mat. Consolidation settlements in the Port Town area (Fig. 4-2) ranged from 1.5-to-2.7 m. Work on the second phase of the reclamation began in 1973 by placing a sand mat over the filled clay, installing wick drains which extended from the lower sand mat at the base of the clay fill (from Phase 1) to the upper sand blanket, and placing an additional 5-to-8 m of sandy fill. Drain wells and vacuum pumps were then used to de-water the lower sand mat. Settlements of about 1.5 m and 5.5 m were recorded at the original mudline and the ground surface, respectively. Somewhat similar methods of accelerating the consolidation process have been implemented in the North Port.

Although the native soils are very similar at the Ports of Kobe and Osaka, the composition of the clayey fill placed in Osaka contrasts with the sandy fill material used in Kobe. The use of the clayey fill appears to have been fortuitous from a seismic perspective, since this soil did not appear to be susceptible to the generation of significant excess pore pressures and significant strength loss under the levels of shaking experienced during the Hyogo-Ken Nanbu Earthquake. However, the clayey fill has posed considerable problems in the operation and maintenance of port facilities due to the continued deformation of the fill leading to differential surface settlements and the distortion of the waterfront quays and crane rails.

### 4.1.3 Quay Walls

The extensive development of the Port of Osaka along shallow river fronts and bay margins, which include calm interior regions as well as bayward sites exposed to significant wave energy, has resulted in the construction of a wide variety of waterfront retaining structures. The port authority has overseen the construction of approximately 100 km of waterfront retaining structures, which are categorized based on configuration and methods of construction as follows:

(a) Shallow Water Berths (total length of about 3.5 km). These consist of anchored steel sheet pile walls, and are employed primarily in the southern portion of the South Port and at numerous mainland facilities.

(b) Deeper Water Berth Structures (total length of about 15.5 km). These structures are used mostly for quay walls for the container terminals at the Inner Port and at the outer South Port. Configurations include anchored interlocking steel pipe piles and concrete caissons.

(c) Inner Port Seawalls (total length of about 60 km). These are predominantly anchored sheet pile walls and pile supported concrete walls.

(d) Revetments (total length of about 23 km). These are double rows of steel sheet piles, sheet pile walls supported by batter steel pipe piles, and concrete caissons. It is noted that concrete caissons comprise most of the revetments in the North Port.

In order to make meaningful comparisons of the seismic performance of various port components, their dates of design and construction can be used as an indicator of the level of seismic design that was used. The post-World War II expansion of the port is evident in the data pertaining to construction of the breakwaters and waterfront retaining structures. A roughly ten year lull in port expansion was ended in 1970 with the initiation of work in the South Port. At that time quay wall construction increased from a rate of roughly 2 km/yr to 6 km/yr. This trend continued to a peak rate of construction of 10 km/yr (1977), which then gradually decreased to 2 km/yr by 1982. The increased development of the port in the early 1970's is similar to that experienced at Kobe Port with the reclamation and construction of facilities at Rokko Island.

Since 1973, concrete caissons have been the most widely used soil retaining and wave dissipation structures in the port (particularly the North Port), although these caissons still comprise a relatively small percentage of the total waterfront structure construction throughout the Port of Osaka. None of the concrete caissons are pile supported, although soil improvement methods have been used extensively to densify foundation soils in the North Port.

The construction sequence that was followed during the development of the South Port involved: (a) excavation of a wide trench in the bayfloor sediments at the location of the proposed wall; (b) backfilling the trench with sand until a flat surface is achieved; (c) driving sheet piles or interlocking pipe piles, while continuing to backfill with sand (anchored bulkheads) or to build up a 4 to 5 m thick mat of rubble fill (caissons), (d) placement of the caisson on the rock fill, and placement of sand backfill adjacent to the caisson by dumping. As previously discussed, the fill used for continued reclamation of eventual backland areas was dredged bayfloor sediments. The soil which had been placed in the foundation and backfill was not densified. Information pertaining to the seismic coefficients used for the design of the retaining structures in the Inner Port and South Port districts was not available during report preparation.

The development of the North Port has posed new design challenges, since municipal waste is used for fill in this reclamation project. The development plan established by the port called for the reclamation of 497 ha with $74 \times 10^6$ m$^3$ of waste between 1972 and the mid-1990's. The waste stream includes dredge spoils, construction debris, household refuse, and industrial waste. In addition to seismic and severe storm concerns, the waterfront retaining structures have also been designed with consideration of the environmental issues associated with offshore waste disposal. The incorporation of geomembrane-lined leachate containment and collection systems along the perimeters of the North Port reclamation islands requires that seismically-induced deformations of the seawalls and partition dikes be minimized. Two common waterfront retaining systems are shown in Figure 4-5. Note that the steel sheet piles (Fig. 4-5a) and the geomembrane (PVC sheet in Fig. 4-5b) are used to minimize the seepage of leachate from the waste fill into Osaka Bay. The adoption by port engineers of substantially improved earthquake-resistant design criteria for waterfront structures in the new North Port region demonstrated an increased awareness of seismic risk to port facilities. The caissons and breakwaters were designed with seismic coefficients of 0.2. In addition, the use of various soil improvement methods for minimizing the potential for seismically-induced ground failures has been extensive.

The methods of ground improvement and the extent to which they have been deployed adjacent to the retaining structures at the North Port are indicated in Figure 4-5. Deep mixing of the native marine clay beneath the concrete caissons has been used to improve the static volume change potential of the soil along the northern perimeter of the South District. Along the southern edge of the North District reclamation, sand compaction piles have been used to increase the liquefaction resistance of the sandy fill. This method of soil improvement has been used along nearly 7.5 km of waterfront in the North Port. Area improvement ratios (defined as the ratio of the area of the improved soil to the total area) are commonly 70% in close proximity to the caissons and decrease to 50% further away. It does not appear that a standard design procedure for specifying the width or depth of the zone of improved soils has been adopted by the port.

Since the early 1970's, the Ports of Kobe and Osaka have carried out major reclamation and development projects at their respective facilities. Although many of the facilities were designed concurrently, significant differences are evident in several aspects of the design and construction of waterfront retaining structures. Factors that would be expected to result in dissimilar seismic performance of the respective port facilities (all other factors being equal) include: (a) seismic resistant design criteria; (b) type of retaining structures; (c) construction and soil improvement methods; and (d) fill materials.

## 4.2   Ground Shaking

The central portion of the Port of Osaka is located approximately 15 km from the vertical projection of the inferred fault rupture for the Hyogo-Ken Nanbu Earthquake. Peak horizontal accelerations experienced during the earthquake in the western portion of the Osaka metropolitan area ranged from approximately 200 cm/sec$^2$ to 270 cm/sec$^2$ (Fig. 2-3 of Chap. 2). The peak accelerations recorded at the Osaka Port Construction Office (in Tempozan Harbor Village) were only 125 cm/sec$^2$ and 178 cm/sec$^2$ in the horizontal directions and 103 cm/sec$^2$ in the vertical component. The lower intensity of shaking at the port offices is somewhat surprising in light of the soft soils that underlie much of this area. It is surmised, based on local distributions of peak ground accelerations in the area (Toki et al., 1995) and the dynamic behavior of soft soils such as those which exist throughout the port, that peak accelerations could have exceeded 0.3 g in the North Port and attenuated to approximately 0.25 g in the South Port. Peak ground surface velocities are estimated from local records to fall in the range of about 25-to-35 cm/sec.

The acceleration time histories at the Port of Osaka were not available at the time this report was prepared; therefore, the duration and frequency content of the motions in this region are not exactly known. The duration of strong shaking is estimated to have been about 5-to-10 seconds. It is clear that the Port of Osaka was subjected to peak horizontal accelerations that were approximately one-half to one-third of the values experienced at the Port of Kobe. However, the ground motions experienced at the Port of Osaka are comparable to or greater than the levels of shaking that have resulted in widespread damage to port facilities in other earthquakes worldwide.

## 4.3   Seismic Performance Overview

Very little damage occurred at the Port of Osaka during the Hyogo-Ken Nanbu Earthquake. Early reports from the Port of Osaka indicated that the only closure of a shipping facility within the complex was the C-9 Terminal located at the northern perimeter of the South Port due to minor crane rail damage (IAPH, 1995). The port remained at full operation following the earthquake, and accepted ships which were rerouted from Kobe. Although ground failures were widespread along the banks of the Yodo River at the northernmost boundary of the port (Fig. 4-6), engineers and other representatives from the Port of Osaka reported that earthquake-induced damage in the port complex was limited to; (a) very minor movement of several seawalls in the Tempozan District (Fig. 4-6), (b) sand boils and ground settlements of up to 30 cm in the Tempozan District, (c) a 20 cm displacement of the deck of the Kizu River Bridge that was quickly repaired, and (d) brief closure of the C-9 Container Terminal. Observations made during an extensive boat tour of the port confirmed, albeit from the bay, the absence of even minor disruption at the waterfront areas of the port (Fig. 4-6).

The exceptional performance of the Port of Osaka facilities is somewhat surprising in light of the levels of ground motion experienced and observations of widespread damage made at numerous modern ports which were subjected to shaking of similar intensity during prior earthquakes. The intensity and duration of ground motions experienced at the Port of Osaka are equal to, or exceed, those experienced at the Ports of San Francisco and Oakland during the 1989 Loma Prieta Earthquake. Widespread liquefaction and pile damage was observed at both of these facilities during the 1989 event. It has been noted that by far the most significant source of earthquake-induced damage to waterfront retaining walls in ports has been pore water pressure buildup in loose to medium dense, saturated, sandy soils that prevail at most port and harbor sites (Werner and Hung, 1982). This conclusion was dramatically confirmed by the extensive damage observed at Kobe Port. The use of the clayey fill throughout the Port of Osaka, the densification of the sandy trench fill beneath the caissons in the North Port, and the short duration of strong shaking appear to have combined to minimize the damage experienced at the Port of Osaka.

Another important aspect of the Port of Osaka is how its excellent seismic performance during the Hyogo-Ken Nanbu Earthquake contrasted with the widespread and extensive damage at the Port of Kobe. It is our view that the Port of Osaka would have experienced damage if its ground motions had been as large as those at the Port of Kobe, although liquefaction effects would have been small due to the predominance of clayey fills at the Osaka Port. Future studies of the contrasting seismic performance of the Osaka and Kobe Ports, particularly at locations of comparable caisson quay walls at each port, will undoubtedly provide important insights into the relative importance of differences in the intensity of the ground shaking, fill materials, construction methods, and degrees of soil improvement at the two ports.

A number of aspects of the seismic performance of the Port of Osaka remain to be investigated, such as: (a) the performance of underground utilities in the Tempozan District where earthquake-induced densification of sandy soils resulted in surface settlements; (b) geotechnical data on the sandy backfill that was dumped from barges behind caissons in the South Port; and (c) the seismic performance of the geosynthetic landfill liners adjacent to retaining structures in the North Port. It is anticipated that additional information from Japan that will be developed in the not to distant future will provide insights into these issues.

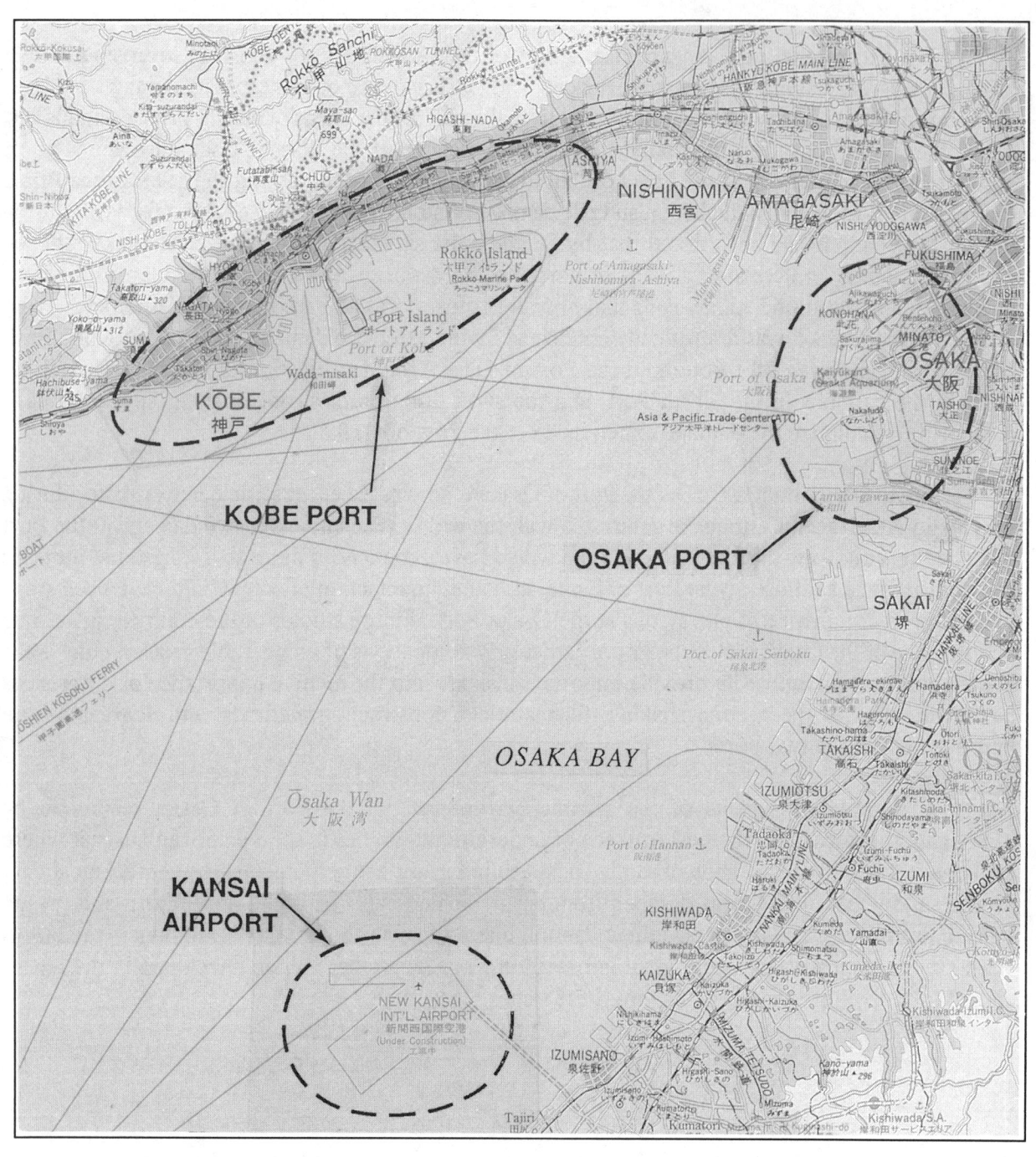

FIGURE 4-1
OSAKA BAY REGION
(EERI, 1995b)

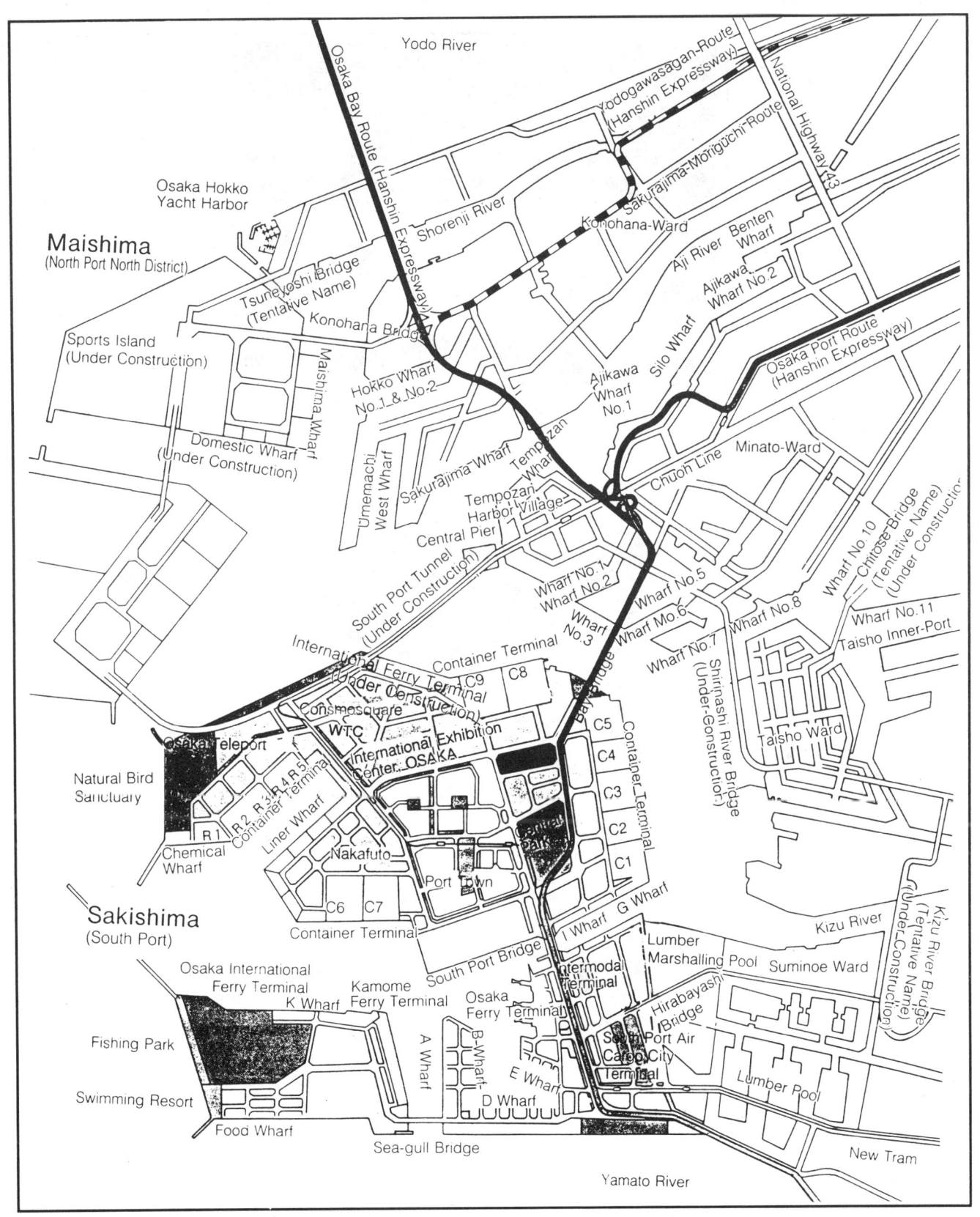

**FIGURE 4-2
PORT OF OSAKA
(Port and Harbor Bureau, 1994)**

a) Tempozan Passenger Terminal

b) South Port - Container Terminal
Berths C8 and C5

FIGURE 4-3
OVERVIEW OF THE PORT OF OSAKA
(Photos from Port and Harbor Bureau, 1994)

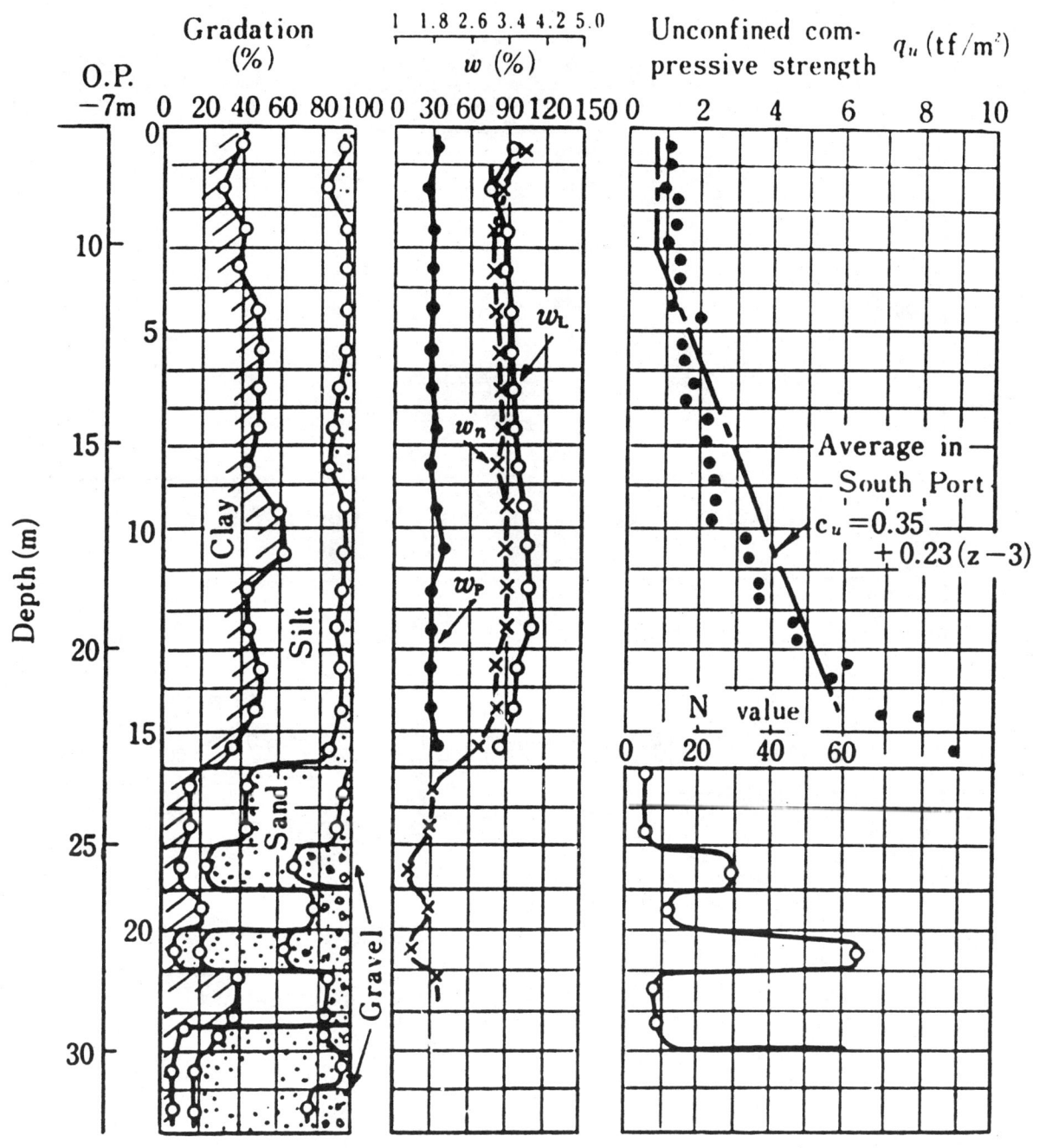

f: volume ratio, w: water content
$w_n$: natural water content, $w_L$: liquid limit and $w_P$: plastic limit

**FIGURE 4-4**
**TYPICAL SOIL PROFILE AT OSAKA PORT**
**(Mikasa and Ohnishi, 1981)**

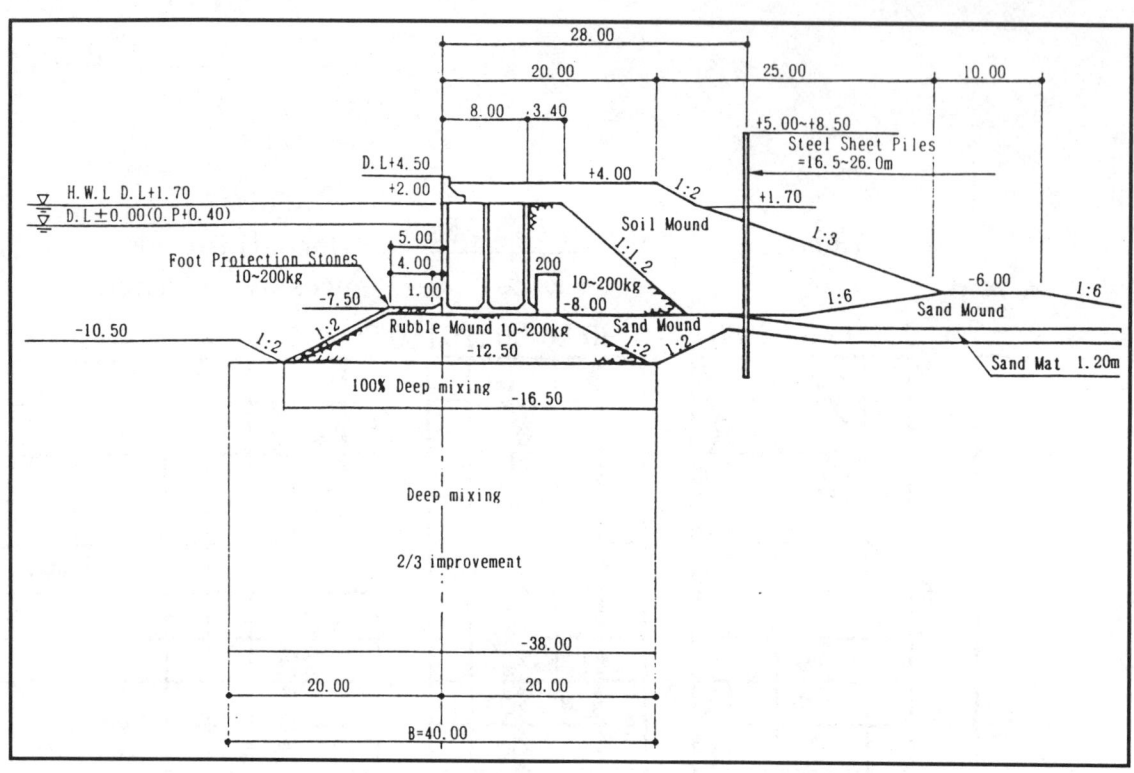

a) Northern Perimeter, South District of North Port

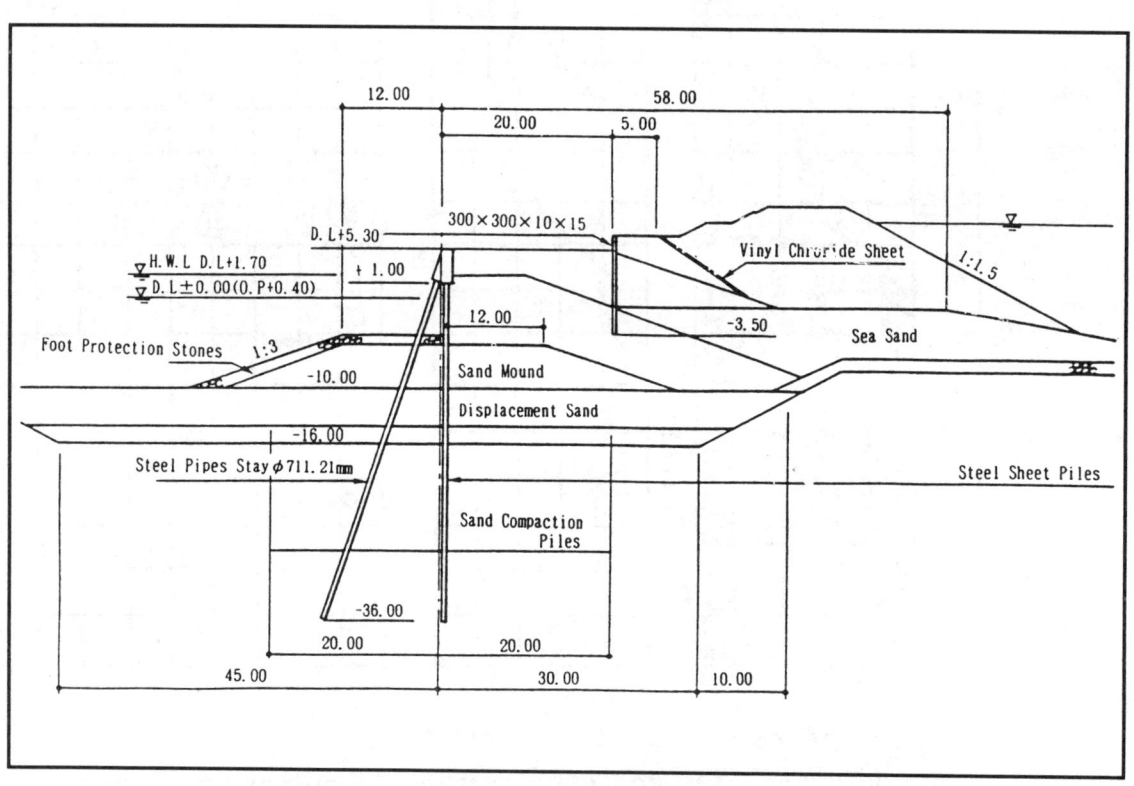

b) Southern Perimeter, North District of North Port

FIGURE 4-5
CROSS SECTIONS - NORTH PORT OF PORT OF OSAKA
(Port and Harbor Bureau, 1990)

a) Liquefaction-induced Ground Failure along Yodo River (Asahi Newspaper Co., 1995a)

b) Evidence of Ground Subsidence Adjacent to a Seawall

c) Good Performance of Facilities at Berth C1 and C2, South Port Container Terminal

d) Berths C6 and C7 of the South Port Public Container Terminal

**FIGURE 4-6
SEISMIC PERFORMANCE OF THE PORT OF OSAKA**

# CHAPTER 5
# KANSAI AIRPORT

## 5.1 General Background

### 5.1.1 Airport Description

The Kansai International Airport is located in the southeastern portion of Osaka Bay, approximately 28 km south of downtown Kobe (Fig. 5-1). It was been built approximately 5 km offshore (to the west of Osaka), on a 510.3 ha island of reclaimed land (Fig. 5-2).

The Kansai International Airport is Japan's first 24-hour airport serving passenger and cargo transport. Its facilities include an extensive passenger terminal complex, administration offices and maintenance facilities, a control tower, and a 3,500 m long runway. Additional infrastructure at the airport site includes the 3.75 km long Sky Gate Bridge (which carries auto and rail traffic), a marine ferry terminal, oil tanker berths, a tank farm for diesel and jet fuel, and a railway station. The airport was opened on September 4, 1994 (KIAC, 1995a).

Officials estimate that this airport will accommodate 160,000 take-offs and landings per year, once its peak capacity is reached. The Kansai International Airport Company (KIAC) has proposed to expand this capacity, by undertaking a 690 ha reclamation project which would increase the size of the main island and would add an additional runway trending west (i.e., further into the bay) from the north corner of the island. This project is scheduled to begin in about 10 years and it will culminate with the development of two additional runways.

### 5.1.2 Airport Island Construction and Settlement Monitoring

Reclamation for the airport island was initiated in 1987. A staged construction sequence was followed in order to allow the 20 m thick layer of soft marine clay to consolidate and gain strength before subsequent layers of fill were placed. Sand drains were used to reduce drainage paths and accelerate the consolidation process. As the perimeter levees approached the surface of the bay, concrete caissons were placed on mats of barge-dumped sand. The foundation soils were not improved before the caissons were placed. Once the caissons were in place, the interior portion of the island was filled. The final depth of the fill at the island is 33 m (KIAC, 1994).

An extensive program of soil improvement was carried out in the near surface fill soils in critical areas of the airport (e.g.; at the control tower, passenger terminal, business offices, and runway). Approximately 320 ha of the 510 ha island received some form of soil improvement. Sand compaction piles were used extensively to improve soils to a depth of 15 m. In addition, the Deep Dynamic Compaction technique was employed in runway areas to densify soils to a depth of approximately 10 m, and a surface vibration (or "tamping") method was used to improve soils in the apron areas to a depth of 7 m. The sandy fill below these depths was not improved.

Construction of the island took place in an average water depth of 18 m. In the five years between 1987 and 1992, $180 \times 10^6$ m$^3$ of sandy fill was brought to the site from borrow areas located throughout the Osaka Bay region. The fill was transported by barges and dumped onto the soft marine sediments that are common in Osaka Bay. KIAC engineers initially anticipated that the surface settlement due to the consolidation of the marine clays underlying the 29 m of fill would be roughly 5.7 m. Observations made at a test fill site indicated that the consolidation settlements would be considerably greater and that the thickness of the fill over the entire island would have to be increased in order to maintain adequate freeboard above storm surge in the bay. An additional 3.5 m of fill (33 m total) was placed and estimates of the settlement due to the fill loading were revised to 11.5 m in 50 years. Continued monitoring of the fill demonstrates that as of early 1995 the settlement had already reached 10 m.

The substantial settlements that are anticipated over the life of the structures at the airport led KIAC engineers to develop structural instrumentation and hydraulic levelling systems that will allow for monitoring and remediation of differential settlements beneath critical facilities. A description of the levelling system used at the air traffic control tower that will occur is provided by Normile (1992). The levelling system employed at the passenger terminal building has been in operation since 1991 and, according to KIAC engineers, it has been activated four times during this time period. In addition to detecting differential settlements due to consolidation, the sensitive fluid-in-tube leveling network is also capable of monitoring settlements due to earthquake-induced densification of the deep fill.

### 5.1.3 Soil Conditions

In general, the soil profile at the Kansai International Airport consists of 33 m of sandy fill underlain by 20 m of soft marine clay, 400 m of interbedded older marine clays and sands, and an alternating sequence of dense sandy soils and hard clay (Fig. 5-3). As discussed above, the near-surface soil is predominantly sandy fill which has been densified in the upper 5-to-15 m. It should be noted that the improved soil is underlain by an extensive layer of unimproved, barge-dumped sandy fill. Geotechnical properties of the improved fill and underlying soils were not available at the time of the preparation of this report.

### 5.1.4 Standard Design of Quay Walls

Several types of retaining structures have been used during the various phases of construction of the island. Small concrete seawalls supported on rock-fill dikes are used along almost all of the 8666 m of the northwest, southwest and southeast margins of the island. A series of sixty-nine steel plate cellular bulkheads (with a diameter of 23 m and a height of 23 m) have been utilized at the southeast corner of the island, adjacent to the Sky Gate Bridge and roadway approach to the airport. The foundation soils beneath these structures were improved with 40 cm diameter, 16 to 20 m long sand compaction piles and, in several areas, deep soil mixing methods were used. In other portions of the island, the marine clays beneath the caissons and dikes were improved with sand compaction piles, sand drains and/or deep soil mixing. This soil improvement is largely confined to the soft clay layers beneath the loose sandy fill. Concrete caissons have been used along the northwest corner of the island, at the location of the marine ferry terminal. These caissons were placed on barge-dumped sandy fill which has not been improved. The concrete caissons were designed with a seismic coefficient of 0.15.

## 5.2 Ground Shaking

The shortest distance from the vertical projection of the Hyogo-Ken Nanbu Earthquake's fault rupture plane to the passenger terminal at the airport is approximately 25-to-27 km. Free-field accelerometers have been deployed at four ground surface locations and one downhole site within the airport, which are: (a) two at the north end of the runway; (b) two at the south end of the runway, and (c) a downhole instrument adjacent to the passenger terminal at the interface between the young marine clay and the older alluvial sediments 50 m below the current ground surface. The peak accelerations recorded by the triaxial instrument at depth were approximately 0.075 g (N-S), 0.10 g (E-W), and 0.10 (V). These ground motions were amplified to values of roughly 0.17 g (N50E), 0.12 g (N40W), and 0.25 (V) at the northern end of the runway; and 0.09 g (N50E), 0.10 g (N40W), and 0.15 g (V) at the southern end of the runway (KIAC, 1995b). The duration of the recorded motions was very short (approximately 5 to 8 sec).

## 5.3 Seismic Performance Overview

At the time of our reconnaissance, KIAC engineers had not detected any damage to either the marine facilities or the airport complex that could be attributed to the earthquake. According to these engineers, evidence of earthquake-induced densification of the fill soils was not apparent from visual inspection of the facilities or from examination of the terminal and control tower levelling systems. In addition, no damage was reported to underground communication or fuel lines, waterfront facilities such as the oil and jet fuel tanker wharf, the pile supported Sky Gate Bridge, the passenger terminal, or the control tower. Very minor lateral ground deformations (as indicated by pavement cracks with widths on the order of 1-to-2 cm) were reported along the perimeter of the island, and earthquake-related ground settlements in the vicinity of the control tower and other major structures were reported to be negligible (Shiraishi, 1995). Similar observations have been reported by Schiff and Wilcoski (1995).

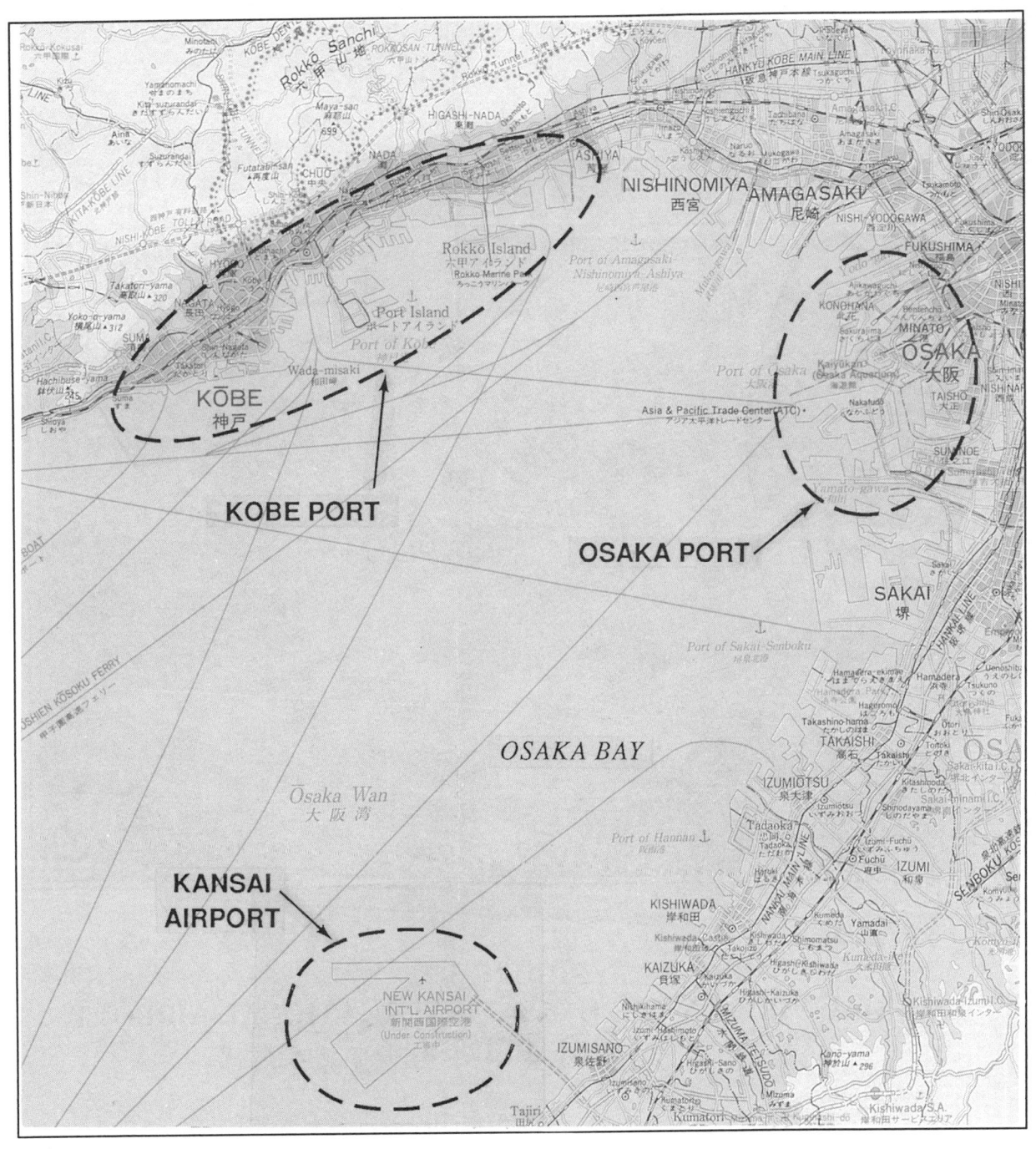

FIGURE 5-1
OSAKA BAY REGION
(EERI, 1995b)

**FIGURE 5-2
AERIAL VIEW OF THE KANSAI INTERNATIONAL AIRPORT
(KIAC, 1995a)**

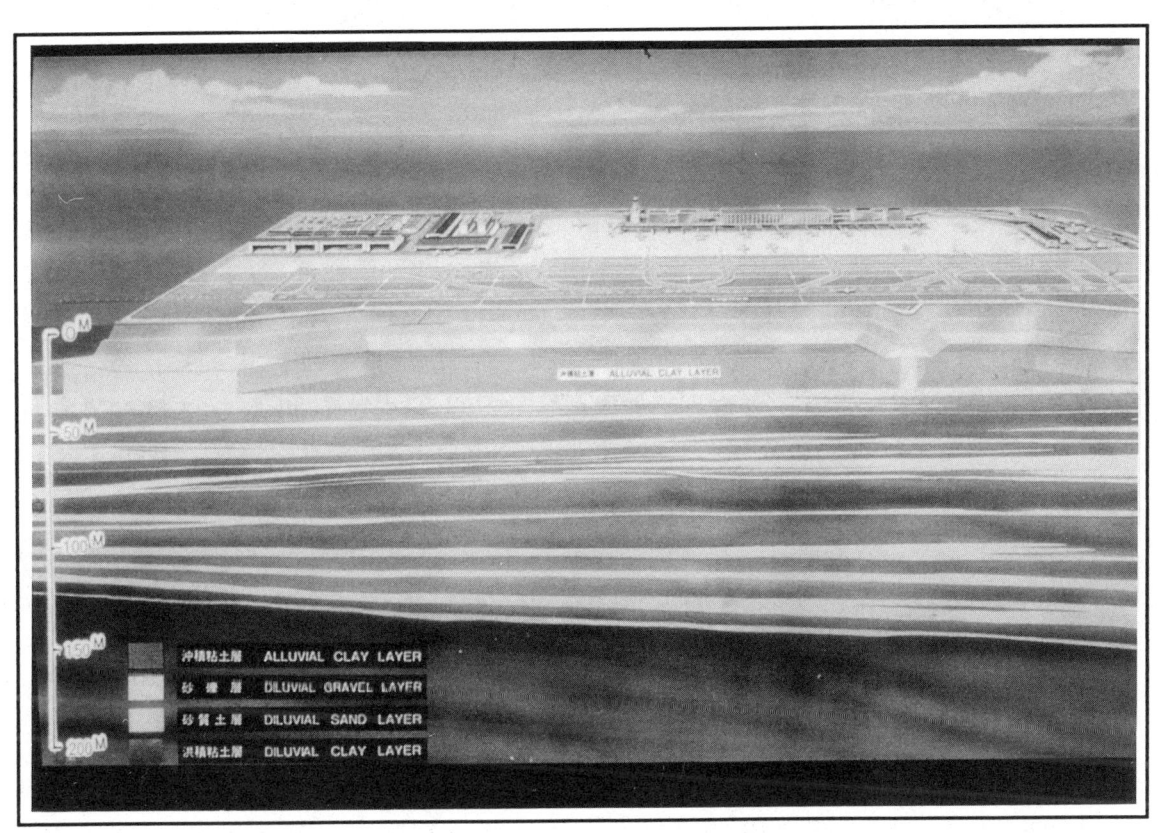

**FIGURE 5-3
ILLUSTRATION OF SOIL CONDITIONS AT
THE KANSAI INTERNATIONAL AIRPORT
(KIAC, 1995a)**

# CHAPTER 6
# CONCLUDING COMMENTS

The Hyogo-Ken-Nanbu Earthquake of January 17, 1995 has provided extensive information regarding the performance of ports and other facilities founded on deep fill (Kansai Airport) under various levels of shaking. From our preliminary assessment of this information, we offer the following perspectives for consideration by the reader:

o   The occurrence of port damage due to liquefaction of fills, as occurred at the Port of Kobe, is not new; rather, as noted in Chapter 1 of this report, liquefaction-induced damage to ports has occurred during many earthquakes. However, what is particularly significant here is the extent and magnitude of the damage that occurred to this major port, and the undoubtedly substantial economic impacts of this damage. Information available at this time indicates estimated port repair costs on the order of $10 billion, and estimated times before the port is fully functional of about two years.

o   The damage to the Port of Kobe is an object lesson as to the vulnerability of ports when subjected to strong, but not excessively long-duration shaking, and when potential seismic risks have not been adequately considered (i.e., no pile-supported quay walls at Port Island and Rokko Island, inadequate compaction of the fills throughout the port, etc.). Clearly, the characteristics of the quay walls, fills, and other key components of ports in regions of the United States with a potential for moderate to severe earthquakes should be compared to those at the Port of Kobe. Based on this comparison, the following questions should be addressed: (a) can significant earthquake damage occur under reasonably likely levels of ground shaking?; (b) if so, what would be the economic consequences of this damage, and are these consequences acceptable?; (c) if these consequences are not acceptable, what level of seismic strengthening should be incorporated, and what are the costs associated with various alternative levels of seismic strengthening?; (d) how do these costs compare with the benefits (i.e., the degree of reduction of damage and associated economic losses) that would result when these alternative levels of seismic strengthening are implemented?

o   The earthquake effects at the Port of Kobe offer examples of both poor performance and good performance of various port components. Clearly, poor performance of inadequately compacted fills and gravity-type caisson quay walls without pile supports was widespread. Also, warehouse buildings comprised of older non-ductile concrete frames or corrugated metal construction with inadequate bracing performed poorly. On the other hand, quay walls with pile supports typically exhibited much better seismic performance, often with relatively little damage and wall movement, even though visible damage to the piles was observed at some locations. Similarly, pile-supported shear wall buildings and/or newer buildings at the port and its islands also typically exhibited good performance, even though the soils around these buildings often settled substantially.

- The seismic design of critical quay walls that are essential to the continued operation of the port should not be based solely on an equivalent lateral force established from a specified lateral force coefficient. Rather, the seismic design of such walls should be based on more advanced procedures that consider the potential effects of pore water pressure buildup in the surrounding soils and fills on the performance of the soil-wall system. Such procedures are now readily available and, in fact, are often used in seismic design/retrofit practice for critical retaining walls at ports and other facilities.

- The Hyogo-Ken Nanbu Earthquake led to widespread damage to bridges in the port area, including bridges that connected Port Island and Rokko Island to the mainland of Kobe. Several ports in the United States also contain key bridge elements that, if damaged, could impact the post-earthquake repair and recovery of the port. The potential seismic risks to ports due to such bridge damage should be considered in the planning of seismic strengthening programs for ports. In addition, the risks to ports from disruption of utility water and natural gas service due to ruptured pipelines should also be considered, since damage to such pipelines was widespread at the Port of Kobe during the earthquake.

- The Port of Osaka represents an important example of very good seismic performance under levels of ground shaking at which numerous other modern ports have experienced widespread damage during past earthquakes. For example, the intensity and duration of ground motions experienced at the Port of Osaka (although moderate when compared to the recorded motions at the Port of Kobe) were equal to or greater than those experienced at the Ports of San Francisco and Oakland during the 1989 Loma Prieta Earthquake, where widespread damage took place. Such factors as the use of the clayey fill throughout the Port of Osaka, and the densification of the sandy trench fill beneath the caissons in the North Port appear to have contributed to the minimal damage experienced at the Port of Osaka during the Hyogo-Ken Nanbu Earthquake.

- Another important aspect of the Port of Osaka is how its excellent seismic performance contrasted with the extensive damage at the Port of Kobe. It is our view that the Port of Osaka would have experienced damage if its ground motions had been as large as those at the Port of Kobe, although liquefaction effects would have been small due to the predominance of clayey fills at the Osaka Port. Future studies of the contrasting seismic performance of the Osaka and Kobe Ports, particularly at locations of comparable caisson quay walls at each port, will undoubtedly provide important insights into the relative importance of differences in the intensity of the ground shaking, fill materials, construction methods, and degrees of soil improvement at the two ports.

- At the levels of shaking experienced at the Kansai Airport (peak horizontal and vertical accelerations of 169 cm/sec$^2$ and 247 cm/sec$^2$ respectively), it appears that the soil improvement methods used in the upper 5-to-15 m of the fill were responsible for the absence of even minor soil deformation. The lack of reported damage to components sensitive to ground deformations -- such as long buildings, pipelines, and communication lines -- attests to the utility of soil improvement methods for seismic stability.

# REFERENCES

Asahi Newspaper Company Ltd. (1995a). Asahigraph, Tokyo, Japan, January.

Asahi Newspaper Company Ltd. (1995b). Weekly AERA, No. 9, Tokyo, Japan, February 25, 82 pp.

Disaster Prevention Research Institute (DPRI) (1995a). Notes on the Hyogo Ken Nanbu Earthquake, Laboratory of Earthquake Prediction: DPRI, Kyoto University, Uji Campus, Kyoto, Japan, 19 pp.

Disaster Prevention Research Institute (DPRI) (1995b). Compilation of Selected Acceleration Time Histories, DPRI, Kyoto University, Uji Campus, Kyoto, Japan.

Earthquake Engineering Research Institute (EERI) (1990). Loma Prieta Earthquake Reconnaissance Report, (Edited by Benuska, K.L.) Supplement to Volume 6 of Earthquake Spectra, Earthquake Engineering Research Institute (EERI) Report 90-01, EERI, Oakland CA, May, 448 pp.

Earthquake Engineering Research Institute (EERI) (1995a). Northridge Earthquake of January 17, 1994 Reconnaissance Report, Volume 1, (Edited by Hall, J.F.), Supplement C to Volume 11 of Earthquake Spectra, Earthquake Engineering Research Institute (EERI) Report 95-03, EERI, Oakland CA, May, 523 pp.

Earthquake Engineering Research Institute (EERI) (1995b). The Hyogo-Ken Nanbu Earthquake, Great Hanshin Earthquake Disaster, January 17, 1995, Preliminary Reconnaissance Report, (Edited by Comartin, C.D., Green, M. and Tubbesing, S.K.), Report 95-04, Oakland, CA, 116 pp.

Eskijian, M.L. (1995). Preliminary Report of the Hyogo Earthquake, Port of Kobe, Japan. State of California Lands Commission, Long Beach CA, February, 10 pp.

Fudo Construction Company (1995). Maps of Soil Improvement on Port Island and Rokko Island. 3 Plates.

Geo-Research Institute (GRI) (1995a). Map of Peak Accelerations recorded throughout the Osaka Bay and Adjacent Regions during the Hyogo Ken Nanbu Earthquake of January, 17, 1995. 1 plate, Osaka Soil Test Laboratory, Osaka, Japan.

Geo-Research Institute (GRI) (1995b). Strong Motion Records from the 1995 Kobe Earthquake, produced in conjunction with the Committee of Earthquake Observation and Research in the Kansai Area (CEORKA), Osaka, Japan.

Griffin, A.C. (1995). The Hyogo-Ken Nanbu Earthquake of January 17, 1995, Follow-Up Report of a Post-Earthquake Investigation of Port and Harbor Facilities in Kobe and Osaka (Draft), ASCE-TCLEE Post-Earthquake Investigation Committee, July 20.

Iai, S., Tsuchida, H. and Koizumi, K. (1989). "A Liquefaction Criterion Based on Field Performances around Seismograph Stations". Soils and Foundations, Vol 29, No. 2, pp. 52-68.

Investigation and Examination Committee for Liquefaction at Osaka Port Facilities (IECLOPF) (1994). Present Conditions of Port Facilities, , Report for the Port of Osaka.

International Association of Ports and Harbors (IAPH). (1995). Interim Report of the State of Ports in Kobe and Osaka, Tokyo, Japan, January 18, 5 pp.

International Center for Disaster-Mitigation Engineering (INCEDE (1995). INCEDE Newsletter: The First 55 Hours, Great Hanshin Earthquake, January 17, 1995, Special Issue, Institute of Industrial Science, University of Tokyo, Tokyo, Japan, January, 20 pp.

Ishihara, K. and Yoshimine, M. (1991). "Evaluation of Settlements in Sand Deposits during Earthquakes", Soils and Foundations, Japanese Society of Soil Mechanics and Foundation Engineering, Volume 32, Number 3, pp 69-80.

Iwasaki, Y. (1995a). Personal Communication with Stephen Dickenson and Stuart Werner. February 22.

Iwasaki, Y. (1995b). Catalogue of Design Specifications for Quay Walls and Pile-Supported Wharves at the Port of Kobe, 9 pp.

Iwasaki, Y., Matsumoto, M. and Yokota, H. (1994), "Innovative Uses of Geophysical Methods- Osaka Deep Basin Structure by S-Wave Reflection", Proceedings of the 13th International Conference on Soil Mechanics and Foundation Engineering, New Delhi, India, pp. 199-200.

Japan National Tourist Organization (JNTO) (1986). Your Guide to Japan., Tokyo, February.

Japan Port and Harbour Association (JPHA) (1989). Technical Standards for Port and Harbour Facilities in Japan (1989 Edition).

Japan Times Ltd. (JTL) (1995). The Great Hanshin Quake, Tokyo, Japan, April, 48 pp.

Kamon, M, Mimura, M. and Katsumi, T. (1995), "Geotechnical Disasters by the 1995 Hyogo-Ken-Nanbu Earthquake", Preliminary Report on the Great Hanshin Earthquake of January 17, 1995, DPRI Newsletter, Special Issue, Disaster Prevention Research Institute, Kyoto University, February, pp 34-40.

Kanai, K. (1983). Engineering Seismology, University of Tokyo Press, Tokyo, Japan, 251 pp.

Kansai International Airport Company, Ltd. (KIAC) (1994). Kansai International Airport, Kansai International Airport Building, Osaka, Japan, May, 16 pp. (in Japanese).

Kansai International Airport Company, Ltd. (KIAC)] (1995a). KIX, The Airport where Dreams Find their Wings, Kansai International Airport, Kansai International Airport Building, Osaka, Japan, 20 pp.

Kansai International Airport Company, Ltd. (KIAC) (1995b). Strong motion records from Kansai International Airport, data provided by Mr. N. Shiraishi, 1st Engineering Dept., KIAC, 2 pp.

Kashima Construction Co. (1995). The 1995 Hyogo-Ken Nanbu Earthquake, Damage Inspection Report, Preliminary Report, Kashima Technical Institute, Japan, February, 186 pp. (in Japanese).

Kawasumi, H. (1951), "Regional Distribution of Earthquake Hazard in Japan", Journal of Architectural Building Science, Vol. 66, pp 773.

Kyodo News Agency (1995). The Great Hanshin Earthquake Disaster, January 17, 1995, Special Issue, Tokyo, Japan, February 4, 90 pp. (in Japanese).

Liftech Consultants, Inc. (1995). Earthquake Damage, Kobe Container Terminals, January 17, 1995. Oakland CA, February 14, 27 pp.

Mainichi Newspaper Co. (1995). Sandei Mainichi, The Great Hanshin Earthquake Disaster, special issue, February 18, Tokyo, Japan, 80 pp. (in Japanese).

Mikasa, M. and Ohnishi, H. (1981). "Soil Improvement by Dewatering in Osaka South Port", Proceedings of Ninth International Conference on Soil Mechanics and Foundation Engineering, Case History Volume, Tokyo, Japan, 1977, pp. 639-657.

Nakakita, Y. and Watanabe, Y. (1981). "Soil Stabilization by Preloading in Kobe Port Island", Proceedings of Ninth International Conference on Soil Mechanics and Foundation Engineering, Case History Volume, Tokyo, Japan, pp. 611-622.

Normile, D. (1992). "Man-Made Island Settles in the Sea", Engineering News Record, Vol. 228, No. 15, The McGraw-Hill Construction Weekly, NY, NY, April, 13, pp. 22-26.

Organizing Committee for Second World Conference on Earthquake Engineering (OCSWCEE) (1960). "Standards of Aseismic Civil Engineering Constructions in Japan", Proceedings of Second World Conference on Earthquake Engineering, Volume on Earthquake Resistant Regulations of the World, Tokyo, Japan, pp. 49-70.

Overseas Coastal Area Development Institute of Japan (OCDI) (1991). Technical Standards for Port and Harbour Facilities in Japan, New Edition, Tokyo, 438 pp.

Port and Harbor Bureau (1990), Port of Osaka and Its Technology; In Commemoration of the 27th International Navigation Congress of PIANC, City of Osaka, May 20-26, 82 pp.

Port and Harbor Bureau (1994), Port of Osaka 1994, City of Osaka, July, 46 pp.

Port and Harbour Research Institute (1995), Plot of Recurrence Interval vs. Peak Ground Acceleration for Kobe Port, Ministry of Transport, Kurihama, Japan.

Port of Kobe (POK) (1994). Port of Kobe, Port and Harbor Bureau, City of Kobe, 37 pp.

Research Group for Active Faults of Japan (RGAFJ) (1991), Active Faults in Japan, Sheet Maps and Inventories, University of Tokyo Press, pp. 272-285.

Schiff, A. and Wilcoski, J. (1995). "Airports", Reconnaissance Report on the 1995 Hyogo-Ken Nanbu Earthquake, American Society of Civil Engineers, Technical Council on Lifeline Earthquake Engineering, (In Preparation).

Seed, H. B. and De Alba, P. (1986). "Use of SPT and CPT Test for Evaluating the Liquefaction Potential of Sands, Proceedings of In-Situ '86: Use of In-Situ Tests in Geotechnical Engineering, American Society of Civil Engineers, Special Publication No. 6, June, pp 281-302.

Shiraishi, N. (1995). Personal Communication, Engineering Division, 1st Engineering Dept., Kansai International Airport Company, Ltd., Kansai International Airport Building, Osaka, Japan, February.

Somerville, P. (1995). "Strong Ground Motions of the Kobe, Japan Earthquake of Jan. 17, 1995, and Implications for Seismic Hazards in California", The Great Hanshin Earthquake Disaster, What Worked and What Didn't?, Proceedings of Structural Engineers Association of Northern California (SEAONC) Spring Seminar Series, SEAONC, San Francisco CA, 14 p.

Sugiyama, Y. (1994), "Neotectonics of Southwest Japan due to the Right-Oblique Subduction of the Philippine Sea Plate", Geofisica Internacional, Vol. 33, No. 1, pp. 53-76.

Toki, K., Irikura, K., and Kagawa, T. (1995). "Strong Motion Data Recorded in Source Area of the Hyogo-Ken Nanbu Earthquake, January 17, 1995" Paper submitted to Journal of Natural Disaster Science, Disaster Prevention Institute, Kyoto University, Uji Campus, Kyoto, Japan, 6 pp. plus figures and tables, (in press).

Tokimatsu, K. and Seed, H.B. (1988). "Evaluation of Settlements in Sands due to Earthquake Shaking", Journal of the Geotechnical Engineering Division, American Society of Civil Engineers, Vol. 113, No. 8, pp 861-878.

Tsuchida, H. (1990), "Japanese Experience with Seismic Design and Response of Port and Harbor Structures", Proc. Worldport LA Seismic Workshop (draft), Port of Los Angeles, CA, 26 pp.

Tsuchida, H. (1995). Personal Communication with Stuart Werner, Stephen Dickenson, and Ian Austin, February 20.

Tsukuda, E. (1987). "Migration of Historical Earthquakes, Central Japan", Proceedings of Conference XXXIX - Directions in Paleoseismology, Albuquerque, NM, U.S. Geological Survey Open-File Report 87-673, pp. 271-284.

Werner, S.D. and Hung, S.J. (1982). Seismic Response of Port and Harbor Facilities, Report No. R-8122-5395, Agbabian Associates, El Segundo CA, Prepared under NSF Grant No. CEE-8012337, 348 pp.

Werner, S.D., Dickenson, S.E., and Erickson, B.P. (1995). "Seismic Guidelines for Ports, A Status Report", PORTS 95, American Society of Civil Engineers, Tampa FLA, 12 pp.

Yomiuri Shimbun (Newspaper) (1995). Photograph from Page 1, January 25, Caption: "Revetment for the Access Road to the Maya Berth has moved a Maximum of 5 m toward the Sea", Photo provided by Prof. M. Hanada, Wasada University.

Youd, T. L. (1995). "Damage due to Liquefaction-Induced Ground Failure, The Great Hanshin Earthquake Disaster", The Great Hanshin Earthquake Disaster - What Worked and What Didn't, Proceedings of Structural Engineers Association of Northern California (SEAONC) Spring Seminar Series, SEAONC, San Francisco CA, 10 pp.

# INDEX

Acceleration: *See* PGA (peak ground acceleration); *See also* Ground motion; Return period
Aftershocks: location of 2-4. *See also* Main shock
Awaji Island: fault system near 2-1, 2-5

Bi-lateral rupture mode 2-1
Bridges, damage to: Dai-Nai Maya Ohashi Bridge 3-16—3-17, 3-58; Hanshin Expressway Bridge 3-17, 3-60; Kizu River Bridge 6-2; Kobe-Ohashi Bridge 3-16, 3-56—3-57; Maya-Ohashi Bridge 3-16—3-17, 3-59; Nadahama-Ohashi Bridge 3-17; rail bridge damage (Kobe) 2-12, 3-17
Buildings, damage to: concrete frame warehouse (Shinko Piers, Kobe) 3-15, 3-50—3-51; corrugated metal warehouse (Naka Wharf, Kobe) 3-15, 3-49; Kobe City (typical) 2-12; wooden warehouse (Hyogo Pier, Kobe) 3-15, 3-49

Caissons. *See* Quays/Quay walls
Clayey fills: deformation of 4-4; resistance to liquefaction 4-3, 4-7, 6-2
Concrete structures, damage to: Concrete frame structures (Shinko Piers, Kobe) 3-15, 3-50—3-51; concrete shear wall structures (Shinko Piers, Kobe) 3-15—3-16, 3-53
Crane damage (Kobe Port) 3-8, 3-38—3-39
Crushed mudstone (Kobe Port) 3-4

Dai-Nai Maya Ohashi Bridge, damage to 3-16—3-17, 3-58
Damage. *See* Earthquake effects
Dickenson, Stephen E. 1-1
Downhole accelerometer array (Port Island, Kobe): array configuration 3-6, 3-32; ground motions at 3-6, 3-33

Earthquake effects: fatalities/injuries 2-2; Kobe city (typical) 2-12; overview 2-2—2-3. *See also* Ground motion; Liquefaction, soil; specific locations (e.g., Kobe, city of; Osaka, Port of, etc.)

Fatalities, earthquake induced 2-2
Fires, typical (Kobe City) 2-12
Fourth Reclamation Area (Kobe Port): soil profiles at 3-3; steel pipe pile seismic performance 3-13, 3-46

Ground motion: duration of (downtown Kobe) 2-1, 2-7; duration of (Kansai International Airport) 5-3; at Kobe Port construction office 3-6, 3-31; long-duration shaking, quay wall vulnerability to 6-1; near Harina Sea 2-3, 2-6; at Port Island downhole array 3-6, 3-33; at Shinko Piers (Kobe Port) 3-15, 3-52. *See also* PGA (peak ground acceleration); Return period

Hanky Ferry Terminal (Kobe Port): quay wall movement and fill settlement 3-36

Hanshin Expressway Bridge (Kobe Port) 3-17, 3-60
Harina Sea: ground motions near 2-3, 2-6
Highway bridges, damage to: Hanshin Expressway Bridge (Kobe Port) 3-17, 3-60; Highway bridge collapse (Kobe City) 2-12
Horizontal velocity: vs. PGA (California, Kobe) 2-2, 2-11
Hyogo Pier (Kobe Port): quay wall movement and fill settlement 3-36; wooden warehouse damage 3-15, 3-49

Injuries, earthquake induced 2-2

Japan, map of (general) 1-3

Kansai International Airport: aerial view 5-5; construction history 5-1—5-2; facility description 5-1; location of (map) 1-4, 5-4; quay walls, design of 5-2—5-3; seismic performance and soil improvements 6-2—6-3; seismic performance (overview) 5-3; soil conditions 5-2, 5-6; structure levelling system 5-2
Kizu River Bridge, displacement of 6-2
Kobe, City of: earthquake damage in (typical) 2-2, 2-12; PGA (downtown accelerometer station) 2-1, 2-7; PGA vs. peak horizontal velocity 2-2, 2-11; PGA vs. return period 2-2, 2-9—2-10
Kobe, Port of: batter piles seismic performance (Maya Piers) 3-12; caisson quay wall sections 3-4, 3-27—3-28; concrete frame warehouse damage (Shinko Piers) 3-15, 3-50—3-51; concrete shear wall damage (Shinko Piers, Kobe) 3-15—3-16, 3-53; corrugated metal warehouse damage (Naka Wharf) 3-15, 3-49; crane damage at 3-8, 3-38—3-39; crushed mudstone, use of 3-4; Dai-Nai Maya Ohashi Bridge damage 3-16—3-17, 3-58; detailed map of 3-21; development chronology 3-1—3-2, 3-19, 3-23; facility description 3-1; grain size vs. liquefaction danger 3-6, 3-24, 3-30; ground motions (at construction office) 3-6, 3-31; ground motions (at Shinko Piers) 3-15, 3-52; ground motions (Port Island downhole array) 3-6, 3-33; Hanshin Expressway Bridge damage 3-17, 3-60; Kobe-Ohashi Bridge damage 3-16, 3-56—3-57; Maya-Ohashi Bridge damage 3-16—3-17, 3-59; Nadahama-Ohashi Bridge damage 3-17; parking structure damage (Shinko Piers) 3-16, 3-54—3-55; pile supported interior structures 3-14—3-15, 3-48; pile supported waterfront structures 3-12—3-14, 3-46—3-47; Port Island downhole array configuration 3-6, 3-32; port location (map) 1-4, 5-4; quay wall movement and fill settlement 3-35—3-37; rail bridge damage 2-12, 3-17; seismic comparison with Osaka Port 6-2; seismic performance overview 3-7; settlement near pile-supported structures 3-14; soil conditions (general discussion) 3-2—3-4; soil improvement (Port Island) 3-3, 3-25; soil

I-1

improvement (Rokko Island) 3-3, 3-26; soil liquefaction, distribution of 3-24; soil profile (Fourth Reclamation Area) 3-3; soil profile (Port Island) 3-2, 3-24; SPCB damage (Maya Piers) 3-9, 3-11—3-12, 3-40, 3-43—3-45; SPCB quay wall configuration details (Maya Piers) 3-9—3-10; steel pipe pile seismic performance 3-13, 3-46—3-48; tanks/tank farm damage 3-17—3-18, 3-62; utility/water line damage 3-17, 3-61; water front retaining structure cross-sections (Maya Pier 1) 3-10, 3-42; wooden warehouse damage (Hyogo Pier) 3-15, 3-49. *See also* Quays/Quay walls

Kobe-Ohashi Bridge: aerial view 3-54; earthquake damage 3-16, 3-56—3-57

Liquefaction, soil: along Yodo River (Osaka Port) 4-13; grain size vs. liquefaction danger 3-6, 3-24, 3-30; at Kobe Port 3-7, 3-24, 3-34; and pore water pressure 2-2, 6-2; sand drains, effectiveness of 3-3; soil liquefaction, distribution of (Kobe Port) 3-24. *See also* Soil conditions

Loma Prieta Earthquake, parallels with 2-1, 4-7, 6-3

Long-duration shaking: quay wall vulnerability to 6-1

Losses, property 2-2

Main shock: location of 2-1, 2-4. *See also* Aftershocks

Maps: Japan (general) 1-3; Kansai International Airport 1-4, 5-4; Kobe Port (detailed map) 3-21; Kobe Port (location) 1-4, 5-4; Osaka Bay region 1-4; Osaka Port (location) 1-4, 5-4; tectonic map, regional 2-5

Maya Piers (Kobe Port): batter piles seismic performance 3-12; layout and aerial view 3-41; quay wall movement and fill settlement 3-35—3-36; SPCB damage 3-9, 3-11—3-12, 3-40, 3-43—3-45; SPCB quay wall configuration details 3-9—3-10; waterfront retaining structure cross-sections 3-10, 3-42

Maya-Ohashi Bridge: earthquake damage (discussion) 3-16—3-17; earthquake damage (photographs) 3-59

Municipal waste as fill (Osaka North Port) 4-5

Naka Pier (Kobe Port): quay wall section 3-28

Naka Wharf (Kobe Port): corrugated metal warehouse damage 3-15, 3-49

North Port (Osaka): ground improvement at 4-5; municipal waste (use as fill) 4-5; reclamation project history 4-2—4-3

Northbridge earthquake, comparison with 2-2

Osaka, Port of: clayey fills, importance of 4-7, 6-2; concrete caissons, use of 4-4; description of facility 4-1; detailed map of 4-9; developmental history of 4-1—4-3; ground shaking (duration, PGA) 4-6, 4-7; ground subsidence, evidence of 4-13; Kizu River Bridge, displacement of 4-6; location of (map) 1-4, 4-8, 5-4; Loma Prieta earthquake, comparison with 4-7; North Port, cross sections of 4-12; North Port ground improvement 4-5; North Port reclamation project 4-2—4-3; quay walls, construction sequence for 4-5; quay walls, types of 4-4; seismic comparison with Kobe Port 4-7, 6-2; seismic performance overview 4-6—4-7, 4-13, 6-2; soil profiles, typical 4-3, 4-11; South Port Container Terminal, views of 4-10, 4-13; South Port reclamation project 4-2, 4-3; Tempozan District, seawall movement at 4-6; Tempozan Passenger Terminal, aerial view of 4-10; Yodo River, liquefaction along 4-13

Osaka Bay region, map of 1-4

Parking structure damage (Shinko Piers, Kobe) 3-16, 3-54—3-55

PGA (peak ground acceleration): accelerometer readings (Port Island downhole array) 3-6, 3-33; distribution of (Kobe area) 2-1, 2-6; at downtown Kobe accelerometer station 2-1, 2-7; horizontal component vector sums 2-1, 2-6; at Kansai International Airport 5-3; largest orthogonal components 2-1, 2-6; near Harina Sea 2-3, 2-6; peak vertical accelerations 2-1; strong motion accelerometer readings (Shinko Piers, Kobe) 3-15, 3-52; vs. peak horizontal velocity (Kobe, California) 2-2, 2-11; vs. return period (Kobe area) 2-2, 2-9—2-10; vs. return period (through out Japan) 2-2, 2-10

Pile supports (seismic performance of): batter piles (Maya Piers, Kobe) 3-12; concrete caisson quay walls (Kobe) 3-7, 3-27; importance of (general comments) 6-1; interior structures (Kobe Port) 3-14—3-15, 3-48; of quay walls (Kobe) 3-4, 3-7, 3-27—3-28; settlement near pile-supported structures 3-14; steel pipe piles (Kobe Port) 3-13, 3-46—3-48; waterfront structures (Kobe) 3-12—3-14, 3-46—3-47. *See also* Quay walls

Pore water pressure: sand drains, effectiveness of 3-3; and seismic design of quay walls 6-2; and soil liquefaction 2-2, 6-2. *See also* Liquefaction, soil; Soil conditions

Port Island (Kobe): concrete shear wall damage 3-15—3-16, 3-53; damaged crane rail (Minami Wharf) 3-39; downhole accelerometer array configuration 3-6, 3-32; ground motions (at downhole accelerometer array) 3-6, 3-33; soil improvement 3-3, 3-25; soil profile 3-2, 3-24

Quays/quay walls: caisson quay wall sections (Kobe) 3-4, 3-27—3-28; caisson seismic design procedure 3-4—3-6; concrete block quay wall performance (Shinko Piers, Kobe) 3-9; concrete caissons, use of (Osaka Port) 4-4; construction sequence (Osaka Port) 4-5; damage at Kobe-

Ohashi Bridge 3-16, 3-56—3-57; design of (Kansai International Airport) 5-2—5-3; liquefaction potential, assessment of 3-8, 3-24; long-duration shaking, vulnerability to 6-1; Naka Pier quay wall section (Kobe) 3-28; performance differences of (Kobe) 3-12; pile supports, importance of 3-7, 6-2; quay wall movement and fill settlement (Kobe) 3-35—3-37; seismic coefficient, calculation of 3-5—3-6, 3-20, 3-29; seismic coefficient and caisson wall performance (Kobe) 3-8; seismic design for pore water pressure 6-2; vibro-compaction, use of (Kobe) 3-3

Rail bridge collapse (Kobe) 2-12
Reclamation, land (Osaka Bay) 2-1, 2-8
Return period: vs. peak acceleration (Kobe area) 2-2, 2-9; vs. peak acceleration (throughout Japan) 2-2, 2-10. *See also* PGA (peak ground acceleration)
Rokko Island: crane damage 3-8, 3-38—3-39; quay wall movement and fill settlement 3-35—3-38; soil improvement on 3-4, 3-26
Rokko Mountains 3-2, 3-4
Rupture mode, bi-lateral 2-1

Sand drains: and soil liquefaction 3-3
Seawall movement/failure: ground subsidence near (Osaka Port) 4-13; Tempozan District (Osaka Port) 4-6
Seismic coefficient, equivalent: and caisson wall performance (Kobe) 3-8; calculation of 3-5—3-6, 3-20, 3-29
Shinko Piers (Kobe Port): concrete block quay wall performance 3-9; concrete frame warehouse damage 3-15, 3-50—3-51; concrete shear wall damage 3-15—3-16, 3-53; crane damage at 3-8, 3-38—3-39; parking structure damage 3-16, 3-54—3-55; quay wall movement and fill settlement 3-36; quay wall section 3-28; strong motion accelerometer readings 3-15, 3-52
Soil conditions: clayey fills, deformation of 4-4; clayey fills, resistance to liquefaction of 4-3—4-4, 6-2; crushed mudstone, use of (Kobe Port) 3-4; Fourth Reclamation Area soil profiles (Kobe Port) 3-3; at Kansai International Airport 5-2, 5-6, 6-2—6-3; North Port cross sections (Osaka Port) 4-12; at Osaka Bay 2-1, 2-8; at Osaka Port 4-3—4-4, 4-11—4-12; at Port Island 3-2—3-3, 3-24—3-25; Rokko Island soil improvement 3-4, 3-26; soil liquefaction, distribution of (Kobe Port) 3-24. *See also* Liquefaction, soil; Pore water pressure
South Port (Osaka): Container Terminal, views of 4-10, 4-13; reclamation project history 4-2, 4-3
SPCBs: *See* Steel plate cellular bulkheads (SPCBs)
Steel plate cellular bulkheads (SPCBs): damage to (Maya Piers, Kobe) 3-9, 3-11—3-12, 3-40, 3-43—3-45; quay wall configuration details (Maya Piers) 3-9—3-10
Strike-slip rupture: main shock, location of 2-1, 2-4; propagation of 2-1
Subway collapse (Kobe City) 2-12

Takahama Wharf: steel pipe piles seismic performance (Kobe Port) 3-13, 3-47
Tanks/tank farms, damage to (Kobe Port) 3-17—3-18, 3-62
TCLEE (Technical Council for Lifeline Earthquake Engineering): Ports and Harbors Committee 1-1
Tectonic map, regional 2-5

Utility/water lines, damage to (Kobe Port) 3-17, 3-61

Water/utility lines, damage to (Kobe Port) 3-17, 3-61
Werner, Stuart D. 1-1

Yodo River, liquefaction along 4-13